Goethe
&
the Power of Rhythm

Johann Wolfgang von Goethe at the age of 79,
by K. J. Stieler

Goethe & the Power of Rhythm

A Biographical Essay

by John Barnes

Adonis Press
Hawthorne Valley
Ghent, New York

second printing, 2002

Published by Adonis Press
320 Route 21C, Ghent, New York 12075

ISBN 0-932776-24-8

Cover design and diagrams by Dale Hushbeck

Printed in the United States of America

This book is dedicated in gratitude
to the four wonderful women in my life:
my mother, Christy MacKaye Barnes,
Frau Anne Schnell,
my first wife, Astrid Clarissa,
and my new wife, Teresa Marie.

Acknowledgment

I was stimulated to explore the 18.6 year rhythm in Goethe's life by Rudolf Meyer's book *Goethes Lebensstufen,* published 50 years ago to commemorate Goethe's 200th birthday. Meyer drew attention at that time to the significance in Goethe's biography of the crises that occur in this rhythm. This essay can be seen as a further elaboration of Meyer's indications.

John Barnes
May 1999

Goethe at the age of 26, by G. M. Kraus

Contents

Rhythm 9

Childhood 11

Adolescence in Leipzig 15

Youth in Strassbourg 19

The Fire of Genius 23

Weimar 26

Italy 41

Weimar Again 48

Friendship with Schiller 51

A New Science of Color 57

Old Age—New Beginnings 68

The Underlying Rhythm in Goethe's Life 79

Notes 92

Bibliography 95

Goethe at the age of 42, by J. H. Lips

Rhythm

As long as you do not attain
From out of death: rebirth!
A gloomy guest you will remain
Upon a dismal earth.[1]
— GOETHE

REMEMBER the mood of a May morning? Sunlight flooding in over the dreaming land, awakening a rising tide of life. Each dewdrop a pearl, flashing green, gold, blue, or crimson as a spear of light irradiates it in sparkling, glistening glory. Each flower, likewise, opening to the sun; each bird joining, full throated, in the music of creation. On such a morning we too can rise up from refreshing sleep filled to brimming with the joyous forces of renewal. Then, like the sun, we pour our forces into the work of the day, until by evening we have spent ourselves and long for the sweet slumber of night, the deep, mysterious ocean of unconsciousness from which we emerge in the morning, reborn. If we fall asleep puzzling over problems or torn with inner pain, we will awaken not only refreshed but clarified and comforted. For, as the Russian fairy tale says, "the morning is wiser than the evening."

Life lives in rhythm. The more actively and consciously we tackle the day, the deeper and more refreshing will be our sleep. The deeper our sleep, the greater our creative powers next day. A pendulum at rest is powerless. The further it swings toward one side, however, the more strongly it will swing toward the other. As it picks up momentum and its arc becomes wider, its power becomes irresistible. Thus the ebb and flow of life's rhythms allow not only for renewal but also for the unfolding of ever greater capacities and powers.

Can we imagine life as a steady, continuous, linear growth process? If we try to, we are soon overwhelmed by a feeling of suffocation: it is as though we were being forced to continue breathing out, evenly, indefinitely.... After a period of growth and expansion, all life reaches a turning point at which a new phase of withdrawal or contraction sets in. Then, however, with a new expansion, new life becomes possible. All life is a breathing: a going out, and a coming back in. Each new breath is a new beginning.

An all-encompassing rhythm is the breathing of the earth through the cycle of the year: in spring we experience an outpouring of life in the growth and unfolding of the plants, in the activity of insects and birds. By midsummer the earth has breathed out fully, and we feel her power unleashed in the rumbling thunderstorms above us. But now the earth inhales, and life pulls in again with the cold autumn winds. By midwinter life has withdrawn into the depths, concentrated itself in bud and seed. Without the death of winter, however, there could be no rebirth in spring. Goethe spoke of death as nature's device to create more life.[2]

Thus death and new life, sleeping and waking, contraction and expansion, in all their endless manifestations, can be seen as polarities between which the healthy, breathing, rhythmical movement of life takes place, ever renewing itself, ever evolving.

Because Goethe's life is extremely well documented through his own very extensive writings and the observations of those who knew him, we are able to form a detailed, many-faceted picture of his long and extraordinarily productive and creative life. A great many books have been written about Goethe, and I am convinced that many more are still waiting to be written.

In this brief essay, however, we will explore only those aspects of his truly archetypal biography which led to his unique development as a scientist. In doing so, we will participate in the powerful rhythm that manifests itself in the dynamic unfolding of his life. It will then become evident how, through rhythmical self-transformation, Goethe was able to attain the new capacities which made him a forerunner of a scientific approach that will, I believe, play an increasingly important role in the 21st century.

Childhood

GOETHE was born as the clock struck twelve noon on August 28th, 1749, in the town of Frankfurt am Main. The sun stood at its highest, and the moon, having just passed the full, stood directly opposed on the other side of the earth. The boy was thought dead at birth, and it was only after considerable exertion on the part of the midwife that he finally drew in his first breath.

Goethe's mother was of a refined patrician family, and her father was mayor of Frankfurt at the time. She had a cheerful, lively disposition. She loved to play and laugh with the children, and she was full of imaginative stories, games, and fun. Goethe, who loved her dearly, always thought of her more as an older sister than as a mother. This relationship may also be partly attributed to the fact that his mother was much younger than his father. Goethe's ancestors on his father's side were of sturdy peasant stock. Through considerable determination and disciplined hard work his father had attained the position of lawyer, and finally, the hand of the mayor's daughter. He was a well-organized, disciplined, thorough, and somewhat stern man who, however, was forward looking, had seen something of the world, and who had made up his mind that his son should have the best education, which he undertook and supervised himself. Goethe's parents complemented each other in every way. They represented archetypal polarities of humanity, which we can see working alternately or together throughout Goethe's life, and the combination of which gave him his balanced universality of being.

> I have my father's stature,
> His ordered, earnest way,
> And mother's cheerful nature,
> Love of stories, fun, and play.[3]

During the course of his early childhood Goethe went through the various childhood diseases each one with its high fever, a transforming trial by fire. This reforging process culminated in a severe bout of smallpox. When convalescence finally set in, and the dry crust of scabby, pockmarked skin fell from his face, leaving the soft new skin unscathed, his features had so changed that an emotional maiden aunt, who had doted on him in his infancy, was unable to overcome her disappointment: gone was the former cherubic harmony, replaced forevermore by a countenance of distinct individuality.

Around about his seventh year this emerging individual sought to establish his own relationship to the divine, creative forces that he sensed so strongly within himself and in the world around him. Intent on erecting an altar upon which to bring a sacrifice to the creator, he went to the upstairs drawing room. Secretly, he selected the most beautiful and precious specimens from his father's well-ordered collection of nature's treasures. As a basis for their arrangement he chose a handsome red-lacquered music stand especially designed for quartets in the form of a four-sided pyramid. Upon this he reverently built up the objects whose tangible wonder and mystery revealed the powers of the unspeakable, intangible object of his devotion. At the tip of the pyramid he placed a small porcelain dish with a piece of incense therein, the sweet-scented glowing of which was to consummate the act of worship.

When the morning sun burst forth from behind the roof of the neighboring house, our little priest was ready, and catching its first bright rays in a magnifying glass, he focused them on the incense, and the festive communion with the world spirit began. The celebration was such a success that the child soon longed to repeat it. This time, however, the little dish could not be found, and the incense was placed directly on the tip of the music stand. Lost in his reverent mood, little Goethe did not notice the golden glow as it ate its way down through the shiny red lacquer, leaving a nasty black spot on the beautiful music stand and putting a final end to the childlike ritual.

Goethe's childhood in Frankfurt was harmonious and protected. The medieval town with its walls and gates, its cobbled streets and squares, was eagerly explored, and its rich diversity and captivating legends fully engaged the boy's lively imagination. Animated social contacts with family and friends, theater and festivals, as well as his ongoing education, all contributed to an active, creative, and secure childhood. Goethe learned easily, even playfully if his lessons were presented in an imaginative way, and when he had reached the age of sixteen, his father and tutors could teach him no more, and he was sent off to the University of Leipzig to study to become a lawyer like his father.

Adolescence In Leipzig

LEIPZIG was known as "klein Paris"—"little Paris": a German center of the brilliant, sophisticated, fashionable culture of the French Enlightenment.

This was the age when young gentlemen appeared with powdered wig, high-heeled, buckled shoes, shapely calves, and cane—women with towering hairdos, deep décolletés, and shapely feet beneath full, hooped dresses. It was the time of geometrically designed gardens and geometrically cut trees, of the minuet and of polished repartee. Geometric abstraction combined with the ornamental excesses of the Rococo style to create a most artificial cultural milieu. It was twenty-four years before the French Revolution; the elaborate, elegant veneer of privileged society was already subject to cutting wit, and the relentless inquiry of youthful reason was grappling with such questions as the natural rights of man. Germany, having only recently recovered from the ravages of the Thirty Years' War, still consisted of some 300 independent feudal states. Politically, socially, and culturally there was much admiration and imitation of the more advanced French culture with its unity, clarity, and elegance.

Goethe arrived in Leipzig with great expectations and took up his studies with diligence. However, he soon found his courses at the university dry, unproductive, and irrelevant to his deep, half-conscious questionings. His lively, searching spirit turned more and more to artistic and social life. He took up drawing and etching, and engaged himself in poetic projects and, above all, in his pursuit of humanity.

As with most adolescent boys, his experience was often deeply painful. For at this age the secure world of childhood has been outgrown, and the strong self-reliance of youth not yet attained. The adolescent is quick to abandon the old authorities and beliefs of childhood but has not yet developed the basis for independent judgment and therefore eagerly identifies with the latest trends.

It was not long before Goethe came to the painful realization that his clothes, which had been especially tailored for him at home, were hopelessly out of fashion. Resolutely, he exchanged his valuable wardrobe for a few stylish suits. More problematic was the fact that his beloved Frankfurt dialect sounded embarrassing in genteel society. Most painful and confusing of all, however, was the cold criticism with which his poetry was received. Goethe had brought all of his poetry with him to Leipzig in the hopes of receiving some praise as well as some helpful criticism. Instead, his work was rejected by the connoisseurs because it failed to comply with the rationalistic and artificial literary principles of the times. Determined to make a new start, Goethe burned his numerous childhood manuscripts on the kitchen stove, alarming his landlady with the clouds of smoke. Most of the poetry of the day was written according to rule and convention and was contrived, sterile, and watered down. Goethe, however, found a model in Lessing, the greatest thinker and writer of the German Enlightenment, a champion of the new ideal of humanity, whose style was direct, concentrated, and to the point.

Even though it contradicted his own deeper nature, Goethe adapted himself with full adolescent devotion and energy to the prevailing culture. Gifted, open, striving, and enterprising as he was, he had soon mastered the elegant lifestyle of Leipzig and had won himself a large circle of loyal friends. It was to these that he was increasingly turning in his pursuit of deeper life experience. Goethe gave himself fully to these human relationships. The wit and humor, the pranks, the tenderness and passion, the creative outbursts, and the despair and exultation born of these friendships can only be indicated here. Goethe's social life became so lively and intense that he neglected to eat or sleep regularly, drank too much beer and coffee, and seriously undermined his health. His moods fluctuated from the heights of euphoria and creative energy to the depths of gloom and despondency. In his autobiography, which he wrote with the wisdom of old age, Goethe speaks of how he drove his organism to the point where its various organs conspired in rebellion in order to save the whole.

One night, in the summer of his third year at Leipzig, Goethe awoke coughing blood. His lung had hemorrhaged, and he experienced a complete physical collapse. For days his life hung by a thread.

With the sudden intervention of this illness, a radical change, a long period of slow convalescence, set in. Convalescence is a "time out" from the pressures and strains of life; it is a withdrawing and turning inward in order to reorganize and rebuild, a time of inner as well as physical re-creation. The delicate yet powerful forces of healing that permeate each cell and organ and bathe the whole physical body in cleansing streams and currents of new life are, at the same time, forces of spiritual renewal, which permeate and bathe the soul, cleansing, rejuvenating, liberating, and opening up new sources of inner strength and inspiration. Illness breaks down hardened, impermeable structures, dissolves all that has grown rigid and disparate. Healing then begins to gently reconstitute, reenliven, and reunite the whole, bringing the physical constitution into harmony with the soul and the spiritual individuality. Convalescence, then, can be thought of as a time of re-formation or rebirth. Goethe writes in his autobiography:

> However, recuperation is always pleasant and comforting, even when it proceeds slowly and wretchedly; and since nature had taken me in hand, I also seemed to have become a different person now, for I had gained a greater cheerfulness of spirit than I had known for a long time, and I was happy to feel inwardly free, even though a protracted illness threatened me outwardly.[4]

Unable to continue his studies, Goethe returned to his home in Frankfurt, where he spent a year and a half resting and recuperating. During this period his inner life was strongly influenced by his friendship with Suzanne von Klettenberg, a woman whose strong, pure religiosity and cheerful disposition made a deep and lasting impression on him. Through her he became acquainted with alchemical traditions. He established a simple laboratory and devoted much time and care to experiments. Some years later he described similar pursuits and mystical experiences such as that of the macrocosm and of the microcosm at the beginning of his *Faust.* He also read Paracelsus, Swedenborg, and Arnold's *History of the Church and Heresy.* This strong turning inward and opening up to the spiritual can only be understood in the light of Goethe's recent closeness to death and of his prolonged convalescence. It was a breakthrough to deeper spiritual sources of his being.

A person of lesser stature might have settled comfortably into the superficial conventions of provincial Rococo culture. To Goethe, however, they belonged to a passing adolescent stage that soon grew constricting and needed to be thrown off to make room for the growth of his own youthful genius. The shedding of the former and the emergence of the latter took place during his illness.

Youth in Strassbourg

IN the spring of his twenty-first year Goethe had recovered his health sufficiently to resume his studies—this time in Strassbourg. The year and a half that he was to spend there may indeed be seen as the springtime of his youth. It was a new turning outward and a meeting of what came toward him with a wellspring of wonderfully fresh, deeply receptive and creative forces.

Soon after arriving, Goethe stood before the Strassbourg cathedral. It made a profound impression on him, which he was not yet able to explain or formulate. He rushed up into the tower. Looking out over the animated city streets as they led toward gardens, orchards, pastures, woods, and distant mountains, a landscape richly formed and cultivated by man, he felt its budding, springtime freshness and beauty as a premonition of the new world of experience that awaited him.

Pursuing his first impression of the cathedral, Goethe gradually came to grasp the overall harmony of its proportions and the lawfulness through which the great diversity and detail of the building only enhanced the unity of the whole. The upper part of its western facade and its one completed tower were not built according to the original plan of Erwin von Steinbach. The second tower was never erected. Out of his sense for the structure of the whole, Goethe sketched the missing tower as he imagined Steinbach to have conceived it. His sketch proved to be remarkably close to the original plan. It was Goethe's imaginative capacity to grasp the architectural principles of the cathedral that later grew into his ability to perceive the living principles of plant formation. In an enthusiastic essay on German architecture, Goethe awakened an appreciation for the Gothic style, which at the time was still considered a monstrous creation of the Dark Ages. While he was studying the cathedral, Goethe successfully overcame his fear of heights, practicing until he was able to stand at ease on a small platform of the tower high above the busy square below.

Thus we see Goethe taking hold of his experience as well as of his own inner development with a self-assurance and rigorousness of which he had been quite incapable during his adolescence in Leipzig.

Goethe soon found out that he would be able to obtain his law degree with a minimum of additional study. He eagerly took the opportunity that this afforded him and signed up for courses in anatomy and chemistry. Most of Goethe's friends were students of medicine, a subject to which he felt strongly drawn, especially because it concerned itself with the whole human being. Here, again, we see Goethe's deep relationship to nature coming to the fore, this time as scientific interest.

Of seminal importance for Goethe's entire subsequent development was his acquaintance and friendship with Herder. Herder was one of the most profound, fecund, and inspired spirits of his generation in central Europe. He was totally antagonistic towards the impotent, superficial rationalism of contemporary scholarship. Herder was one of the first to conceive of the idea of evolution. In contrast to Darwin's later view, however, he saw evolution as proceeding from a wellspring of divine creativity whose intentions were realized on ever-higher levels and whose work finally culminated in the evolution of the human being.

To Herder, these creative forces revealed themselves most gloriously in the childhood of humanity, in the ancient cultures in which human beings still experienced the working of the divine in nature. Contrary to contemporary conjecture, Herder was convinced that the origin of human language lay in inspired poetic participation in divine creativity itself.

Goethe visited Herder in the Hotel "zum Geist" ("to the spirit") where he was staying in a darkened room while being treated for an eye ailment. It was here that Herder introduced Goethe to the divinely inspired world of Homer, to the deep wisdom of the Old Testament, and above all, to the genius of Shakespeare. Shakespeare was still virtually unknown in Germany at the time, and Herder extolled his work, contrasting it with the artificial convention of French drama and calling him not a poet but a "creator" who, in collaboration with the world spirit itself, had brought forth "the most alive and comprehensive depiction of human nature."[5]

Herder's powerful intuitive insights resonated deeply in Goethe. They allowed him to see into the creative heart of nature and humanity and provided fertile ground for his own spiritual growth and for the unfolding of his own creative genius. Herder broke the ground not only for Goethe but for a new epoch. The later romantic movement in Germany also traced its origins back to him.

Spurred partly by Herder's exhortations to seek out the uncorrupted genius still living in the folk culture, driven also by the wish to further enhance his growing health and strength, by his youthful love of adventure and by his joy in the rich variety and beauty of nature, Goethe made many riding excursions into the countryside around Strassbourg.

On one of these expeditions he made the acquaintance of Friederike Brion, the daughter of a village pastor. In a fully unsentimental way she was like a beautiful wildflower: the charm of the whole folk culture and of its natural environment, including the heavens, seemed to find expression in her graceful form and manner. Nothing artificial masked the cheerful, confiding openness of her being. Finding in her the embodiment of all he sought and longed for, Goethe was irresistibly drawn to her. Lifted by love into a higher sphere, life took on for them a greater intensity and power. A May morning now became a celebration of bursting, sparkling life and love. Goethe's lyric poetry of the time is unsurpassed in its sheer direct power and purity of feeling.

After some time, however, Goethe withdrew. He realized that the wildflower could not be transplanted to the city, where it would wilt among the hybrids. Beautiful as his relationship with Friederike was, he sensed that it could not be sustained. But he was deeply shattered by the pain that he had caused the one who would live out the rest of her life in the memory of this golden, youthful love.

The Fire of Genius

AFTER finishing his studies, Goethe moved back to Frankfurt. He took to wandering through the land, by himself, pondering the power that had led him to so deeply injure the dearest of beings. He felt this irrepressible, irresistible power like a force of nature leading, guiding, inspiring him, carrying him as on mighty wings through the depths and heights of human experience. Thus wandering towards an approaching storm, he muttered a hymn to his Genius: a power greater than himself, of higher origin, which endowed him with glowing, indomitable courage and boundless creativity. This and other poetry that he wrote at the time, such as "Prometheus" or the "Urfaust," pay no heed to poetic form or convention: they are spontaneous, passionate outbursts, directly inspired by his Genius.

Already in Strassbourg Goethe had become the leader of a revolutionary generation of writers who, inspired by Herder and Rousseau, by the heroic individualism of the germanic past, by a disdain for the effete sophistication of the corrupt nobility, and above all, by the fire of their own youthful inspiration, swept aside all the empty formalism of the preceding age. The impetuous vitality of genius, the power of honest feeling, the love of freedom: these were chief characteristics of this movement later called "Storm and Stress." It is perhaps best summed up as a revolution of the heart and will against the predominance of the cold, detached intellect.

Goethe, in his early twenties, was the natural leader of this movement. His outer physical appearance fully conveyed the power of his inner capacities. He was of medium stature, his torso and head relatively large but well proportioned. His features were strong and harmonious. His large, observant, dark eyes mirrored the depth, intensity, and richness of his inner life. He was a good rider, dancer, and ice skater, a fascinating story teller, a fine draftsman, a poet who could compose verse at the drop of a hat and whose inspired poetry ranked with the greatest in the world. Young Goethe was an open, tolerant, sociable person whose sheer human genius naturally made him the center of any gathering and won the hearts of all who shared his creative and deeply human disposition. There were few women who could help falling in love with him.

Goethe's first national fame came with his powerful, youthful play "Götz von Berlichingen," which depicts the tragic downfall of an unsophisticated moral giant at the hands of the legalistic scheming of shallow moral weaklings.

Sudden European fame followed soon upon this first success: While in Wetzlar to gain legal experience, Goethe fell in love with the bride of a good friend and colleague. Unable to fully conceal or control his violent passions, indeed more and more possessed by his unrequited love, Goethe finally tore himself away and left the town. Not long afterwards Goethe heard of the suicide of a young man in similar circumstances. This tragic story, so close to his own, galvanized Goethe's imagination. Now able to pour his innermost, sublime yet desperate experience into an objective outer form—thus freeing himself of the unbearable turmoil dammed up within him—he chose the genre of a novel written in the form of letters and diary entries, a form well suited to contain the flood of intimate feelings. This correspondence, published under the title of *The Sufferings of*

Young Werther, is the deeply moving confession of a highly sensitive, loving soul whose soaring passion grows ever more absolute and all consuming and ever more glaringly and hopelessly opposed to reality. The discrepancy between Werther's subjective feeling, which fluctuates between euphoria and heart-wrenching despair, and the sober world that runs its course without him, becomes a gaping abyss and leads ultimately to Werther's suicide.

The effect of the novel was unprecedented. "Werther fever" swept through the hearts of Europe. Few were able to read the book without shedding tears profusely on its pages. After generations of suppressed emotion, Europe reveled in the free flow of feeling. Enthusiastic young men wore Werther's yellow pants, and public displays of emotion became the fashion. Goethe had become the "superstar" of a new era.

Yet we must not forget that, in writing "Werther," Goethe had objectified and overcome the Werther in himself, thus healing himself of the very "fever" he had inoculated into his contemporaries.

Weimar—From Poetry to Science

IT was just at this point in his life—when he had achieved immense popularity and his soul-pendulum had swung powerfully, indeed dangerously, dizzyingly high into the realm of subjective experience—that Goethe took his first independent step into life, a step that was to lead him back—onwards and also downwards as it were—into the limitations of the "objective" world. In 1775, at the age of 26, Goethe accepted an invitation to become the companion of the eighteen-year-old Duke of Weimar and eventually take on ministerial duties at his court.

Goethe, who was nearing the end of his "Storm and Stress" period, hit it off well with the wild young Duke, who was just beginning his. Together they stormed through the Dukedom, hunting on horseback through the peasants' fields, relishing their girls and wine at rollicking village dances and often turning even the sedate court itself into a riotous carnival scene. Gradually and imperceptibly, however, Goethe began to turn the young Duke's attention toward his duties as regent. More and more, also, he was drawn into the inner circles of the court and government. As a member of the Privy Council he first was appointed Minister of Public Works—which meant overseeing the mines and, later, the roads of the region. He was then also made Minister of War—which meant recruiting, arming, and maintaining an army of 660 men. Finally, after seven years, he was appointed Finance Minister and President of the Council. He was now responsible for the court, its theater, public buildings and parks, roads and footpaths, mines, industry, police, military, schools, forestry, and finances! In effect, Goethe now carried the main responsibility for the Dukedom.

The pendulum had swung a full 180 degrees! The same Goethe who at twenty-five had written "Werther," who had created out of his own experience that passionate champion of sublime self-centeredness whose rudderless life-ship, driven before the tempestuous winds of his emotions, is wrecked on the rocky cliffs of reality—this same Goethe was now, at thirty-three, the highest official at the starchy Court of Weimar! We find him in the mines, visiting wool factories, trying a new method of selling timber, establishing a laboratory, having a castle renovated, an old hospital demolished, the arch of a bridge widened, creating a marble mill in a brick factory, reckoning the profit in a favorable season of the Ducal grain after deducting all the demands of the court, the servants, and the military. Goethe tackled these tasks with rigorousness and perseverance and proved himself a capable, far-sighted, and realistic administrator. When there was a serious emergency, Goethe responded unreservedly. At the terrible fire in Apolda and the catastrophic flood in Jena, he was at the center of the rescue effort, grasping and directing the situation and laying on hand in a way unusual for a high official.

In this way Goethe gained all manner of practical and human experience, some of which was exhilarating and enriching, much of which, however, was tedious and frustrating for a creative, sensitive, and idealistic young soul to bear. The fiery young poet would hardly have been able to endure the dry drudgery of the paper work, the pedantry of the bureaucracy, or the stiff etiquette of the court had it not been for his relationship with Charlotte von Stein. She was seven years older than Goethe, married, inwardly steadfast and noble and able to keep the deep love that drew them together on a beautiful, free, soul-and-spiritual level. He came to her often in the evening in an inner storm, deeply frustrated by the events of the day, looking for the solace, peace, and inner rejuvenation that only she could give him. She helped him to overcome his youthful impatience, to transform his self-centered emotions into purified perceptions of the other, to accept the limitations and constrictions that life imposed upon him, and in so doing, to gain self-mastery and inner freedom. Until he found his own higher self in Italy, she was, as Friedrich Hiebel has pointed out, Goethe's guiding star—a steady, soothing, higher self to whom he confided his innermost thoughts almost on a daily basis.[6] During the ten years between Goethe's arrival in Weimar and his departure for Italy, she was "the eternal feminine" which drew him onward.[7]

Goethe's relationship with Frau von Stein demanded chastity and renunciation. In a sense Goethe's administrative duties, which consumed much of his time, also meant renunciation—namely with regard to his own creative productivity. During the ten-year period we are now concerned with, Goethe's literary productivity declined steadily, and its creative source seemed to be drying up. His publishers gave him up as a youthful genius who had lost his creativity.

Renunciation, self-control, and the focusing of inner forces on the diverse demands imposed by life, abstaining from the outer consummation and fulfillment of a love affair and of a youthful literary career—all this can also be seen in a positive light. For if Goethe had remained the unrestricted youthful genius and had pursued his subjective desires and inclinations, he might have extended his popularity for a few more years, but he would never have attained the objective depth of his classical period. These first ten years in Weimar were a necessary schooling and precondition for his later greatness. Through a certain asceticism regarding himself, and by concentrating on the demands of the outer world, Goethe moved powerfully towards the pole of life which we may characterize as the objective.

This swing in Goethe's development from the subjective to the objective pole is reflected in his novel *William Meister's Apprenticeship,* in which he depicts the hero's transition from adolescence through youth to manhood. Because Goethe possessed the predisposition and inclination to keep a certain distance from his life experience, he was able to take it up in his imagination, and penetrate and objectify it until it revealed its essential nature. All of his works are basically the ripe fruits of his own experience. It was by learning from life in this way that he gradually developed the great, practical life wisdom for which he is famous.

However, these Weimar years with their movement toward the objective led inevitably to a crisis, which, in turn, led to Goethe's great breakthrough into a new realm of objective creativity.

This movement, crisis, and breakthrough are clearly reflected in the progress of Goethe's scientific studies at the time.

> However, I actually only entered into practical life and into the sphere of science when I was happily accepted into the Weimar circle. Immediately upon my arrival I experienced the inestimable benefit of exchanging the atmosphere of chamber and city with country, forest, and garden air.[8]

Goethe's new surroundings and activities in Weimar—rather than appearing foreign and inimical to him—seemed to meet an inner need and longing. He took to this new world with an all-embracing, seemingly unquenchable interest—like a babe to its mother's breast. When he moved into his garden house in Weimar, his interest in botany began in earnest. Enthralled with the delicate and dramatic changes in the weather, his abilities as a meteorologist developed over the years to the point where, with an observant glance at the sky and at the barometer, he could accurately predict the weather for the next twenty-four hours. The anatomical studies that he had begun as a student in Strassbourg now reawakened with an intensity fired by new and deeper insight. Goethe's work in the mines led to a passionate, if quiet, interest in mineralogy and geology. This leads us to one of the central riddles of Goethe's being: why should an artist, a poet who is fully at home in the world of the human soul and in the seemingly ephemeral realm of free imagination, be so irresistibly drawn to the world of natural science, yes even to what one might think of as the antithesis of poetic creativity: the seemingly unchanging and speechless world of the stones? In his essay "On Granite" Goethe characterizes the deep feeling that one can have on the bald summit of a granite mountain. His interest in the extraordinary consistency and variety of this most ancient of all minerals may seem, he realizes, to some of his readers to be a paradox, which he then strives to explain:

> And thus anyone who knows the fascination natural mysteries hold for man will understand why I have departed from my usual realm of observation and turned with passionate fervor to this one. I do not fear the accusation that a contrary spirit has led me away from my consideration and depiction of the human heart, the youngest, most diverse, most fluid, most changeable, most vulnerable part of creation, and has brought me to the observation of the oldest, firmest, deepest, most unshakable son of nature. It is evident that all things in nature have a clear relationship to one another, and that the questing spirit resists being denied what it can attain. I have suffered and continue to suffer much through the inconstancy of human opinion, through its sudden changes in me and in others, and I may be forgiven my desire for that sublime tranquility which surrounds us when we stand in the solitude and silence of nature, vast and eloquent with its still voice. Let those who are aware of this feeling follow me on my journey.

We are reminded of the boy who placed crystals and minerals on his father's music stand:

> Here, on this primal and everlasting altar raised directly on the ground of creation, I bring the being of all beings a sacrifice. I feel the first and most abiding origin of our existence; I survey the world with its undulating valleys and its distant fruitful meadows, my soul is exalted beyond itself and above all the world, and it yearns for the heavens, which are so near.

True to his nature, Goethe does not lose himself in the contemplation of the eternal:

> But soon the burning sun will bring back thirst and hunger, the human necessities. Our observer's gaze will seek out the very valleys over which his spirit had soared. He will envy the dwellers in those more abundant and plentifully watered plains, the inhabitants who have built their happy homes on the debris and ruin of error and opinion, who scratch in the dust of their ancestors and quietly meet the modest needs of their daily existence within those narrow confines.[9]

From the eternal, the pendulum of his soul sweeps back to the confines of earthly life. Here we have an example of the breathing activity that characterized Goethe's inner life as well as his life as a whole.

The question of the relationship between Goethe the poet and Goethe the scientist leads us to an ever-deeper understanding of Goethe's essential nature. Goethe was not an artist for art's sake. His poetic works are nothing other than an attempt to penetrate, digest, and objectify his own experience, to form it in such a way that the essential, the inner being or reality, shines through it and becomes manifest. He would leave a poetic work unfinished rather than force it to a conclusion that did not arise organically and truly out of the subject matter itself. As a poet or novelist, Goethe was able to slip into the soul experience of a human being and portray it in such a way that it became a deeply moving, yet comprehensible and true revelation of an aspect of that person's essential being. He was, therefore, only following his own deepest inclination when he strove to enter into the natural world as well in such a way that its inner lawfulness, its essential nature, revealed itself to him. It is in this sense that Goethe has often been called a "seer." His central striving united realms that we today conceive of as separate: art, science, and religion. Already in his early twenties, in Strassbourg, Goethe formulated with youthful directness the central striving of Faust—which was, indeed, his own—in two famous lines:

> Dass ich erkenne was die Welt
> Im Innersten zusammenhält
> [That I might see what secret force
> Hides in the world and rules its course][10]

Goethe the poet, Goethe the scientist, and Goethe the religious seeker all grow out of this central striving.

In his later years Goethe sometimes called the object of his Faustian striving "Gott-Natur"—"the divine in nature."

This divine, creative inner being of nature is the subject of his early essay "Nature," in which we can feel the powerful inspiration of his youthful genius:

> Nature! We are surrounded and embraced by her—powerless to leave her and powerless to enter her more deeply. Unasked and without warning she sweeps us away in the round of her dance and dances on until we fall exhausted from her arms.
>
> She brings forth ever new forms: what is there, never was; what was, never will return. All is new, and yet forever old.
>
> We live within her, and are strangers to her. She speaks perpetually with us, and does not betray her secret. We work on her constantly, and yet have no power over her.
>
> All her effort seems bent toward individuality, and she cares nothing for individuals. She builds always, destroys always, and her workshop is beyond our reach. She lives in countless children, and the mother—where is she? She is the sole artist, creating extreme contrast out of the simplest material, the greatest perfection seemingly without effort, the most definite clarity always veiled with a touch of softness. Each of her works has its own being, each of her phenomena its separate idea, and yet all create a single whole.[11]

This rhapsodic essay, only the beginning of which can be quoted here, reveals the dynamic quality of Goethe's holistic, monistic world view and characterizes the deeper impulse of his spiritual striving. It is an expression of this impulse in its youthful, inspired form before it had led to concrete scientific insights.

It was now in Weimar, however, that Goethe set himself the task of coming to grips with nature in a fully scientific way. His first major achievement as a scientist, in the field of comparative anatomy, was the discovery of the premaxillary bones in the human skull. Until that time, those who wanted to demonstrate that humanity constituted a separate kingdom from the animals based their arguments in part on the absence of the premaxillary bones in the upper jaw of the human skull. These bones, which are highly developed in animals with protruding muzzles, such as the horse, have become so rudimentary in man that they can only be identified with difficulty. Goethe's discovery was a deeply satisfying confirmation of his intuitive conception of nature, for it shows nature's consistency in creating variety through the transformation, or metamorphosis, of the same basic plan or "type." Thus all vertebrates, including the human being, are built according to the same flexible plan, each species manifesting it in a fixed and very specific way. For Goethe it was grotesque to want to distinguish between the human and the animal kingdom through such an apparent minor inconsistency as the absence of a pair of little bones. For him the essential difference lay in the overall harmony of the type's manifestation in the human being—a harmony and balance which allow for the free unfolding of creative, moral and spiritual capacities.

In his study of plants Goethe was guided by the system atic work of the Swedish botanist Linnaeus. Linnaeus had laid the groundwork for a comprehensive classification of the entire plant kingdom. He had endeavored to distinguish as exactly as possible the various families, species, and subspecies. The nature of his undertaking forced him to divide and subdivide his categories to create order within the almost limitless variety of the plant world. Goethe benefited greatly from Linnaeus's work. However, he also grew increasingly aware of its limitations. These he encountered when he observed the same plant species growing under different conditions in different environments. The same plant species grows differently at the edge of a forest than it does a few yards into its dark interior; it manifests quite differently on the seashore than on a high mountain slope. The more Goethe observed, the more he became aware of the inner mobility of the plant and the more difficult it became for him to establish a definite dividing line between one subspecies and another. In short, Goethe soon felt uncomfortable with Linnaeus's static system, which was constantly distinguishing, defining, and separating and which was unable to move with nature from one form to another. Goethe sensed that there was unity in the endless multiplicity of the plant world.

In studying an individual plant, also, Goethe could not be satisfied with simply labeling its various parts. For example, in the poinsettia, which we like to buy at Christmas time, the bright red leaves appear like petals. Yet are they not really the same as the green leaves below, except that, through their color, they have been integrated into the flower? Goethe was constantly finding such irregular transitions, which seemed to call for a more fluid, flexible way of thinking. He sensed that the plant, which is constantly in a state of becoming, ever transforming itself, cannot be grasped in its essential nature by a thinking that is static, passive, and only able to separate, define, and identify the phenomena. He sensed that there is in each plant a living formative principle that brings forth the sequence of forms in its life cycle.

The groping search for a unifying formative principle in the mobile, multiform plant kingdom engaged Goethe's deepest cognitive forces in a tremendous struggle, for it meant nothing less than the development of a qualitatively new kind of thinking. Burdened by his administrative duties and unable to devote himself to his botanical studies, Goethe reached a cognitive impasse that was deeply bound up with the existential crisis he was experiencing at the time.

This crisis came to a head in Goethe's tenth year in Weimar, when he was thirty-seven years old. It was his midlife crisis. However, in Goethe's case, it took on dimensions of far-reaching importance. We all experience around this age the final ebbing of our youthful forces. This often leads to a crisis in marital or professional life. Unless we are able, out of our own inner activity, to find new sources of inspiration and creativity, we simply stop evolving and gradually become the slave of our habits.

Youthful creativity is inspired by the enthusiasm of youthful genius, which we bring with us at birth, which surrounds us in childhood, draws into us in our youth, and gradually dies away as we approach middle age. It is the heavenly force that Wordsworth so well described in his "Ode on Intimations of Immortality, from Recollections of Early Childhood:"

> Our birth is but a sleep and a forgetting:
> The Soul that rises with us, our life's Star,
> Hath had elsewhere its setting,
> And cometh from afar:
> Not in entire forgetfulness
> And not in utter nakedness,
> But trailing clouds of glory do we come
> From God, who is our home:
> Heaven lies about us in our infancy!
> Shades of the prison-house begin to close
> Upon the growing Boy,
> But he beholds the light, and whence it flows,
> He sees it in his joy;
> The Youth, who daily farther from the east
> Must travel, still is Nature's priest,
> And by the vision splendid
> Is on his way attended;
> At length the Man perceives it die away,
> And fade into the light of common day.[12]

This is the source that the romantics drew from—a "subjective" spiritual force that fades the further it enters into the "objective" realities of life. It is no accident that the creative period in the lives of the romantics was their youth and that most of them died before their mid-thirties. It is of tremendous significance that Goethe was able to go through the death of these forces and to successfully attain new creative spiritual capacities. For the importance of Goethe for us today lies not so much in the powerfully inspired works of his youth, as in the momentous new beginning that he was able to make at midlife. Those who see Goethe as a promising, revolutionary young poet who succumbed to a retiring and unproductive aristocratic existence at the court of Weimar fail to see and understand the epoch-making spiritual evolution that Goethe achieved in the second half of his life.

By the year 1786 Goethe could stand it in Weimar no longer. The weight of his administrative duties had become an unbearable burden. His literary work was stagnating. The life at court was stifling for him, and he was often silent, introspective, and forbidding, even morbid in manner. His features were growing haggard. He felt himself close to a breakthrough in botany, but he lacked the time and energy to pursue his hunches. He was experiencing what might be described as death by soul suffocation.

Yet Goethe was not helpless. He had a deep, intuitive sense for what he needed. Secretly, he planned his flight to Italy.

Italy

TELLING only the Duke of his intentions, Goethe disappeared after celebrating his thirty-seventh birthday in Carlsbad. Traveling incognito by mail coach, he was on his way toward the Alps—and then down through Italy to Rome! With every mile that drew him closer to his destination, a weight seemed to lift from his soul. Each new impression that welcomed him on his way he drank in deeply, and it engendered in him a font of new life. He began to breathe more and more deeply. Entering Italy, he experienced a deep inner renewal. The clarity that pervades this southern country had a profound, healing effect upon his soul.

In December he wrote repeatedly in Rome of his experience of rebirth, of radical transformation, of a renewal right down "to the very marrow of my bones."[13]

Goethe's inner renewal was accompanied by deepening insights into plant morphology. Passing over the Alps, he saw familiar plant species adapt to higher altitudes. After visiting the botanical garden in Padua, he writes on September 27th:

> It is agreeable and instructive to wander amidst vegetation that is foreign to us. We eventually think no more at all about plants we are accustomed to, like other long familiar objects; and what is observation without thought? Here in this newly encountered diversity that idea of mine keeps gaining strength, namely, that perhaps all plant forms can be derived from one.[14]

On February 19th, 1787, he writes in Rome that he is on his way to discovering how nature "develops the manifold out of the simple." On March 25th, on a contemplative walk on the beach near Naples, an illuminating thought brings him closer to a grasp of the archetypal plant.

Finally, on May 17th, he writes in a letter to Herder:

> Moreover, I must tell you confidentially that I am very close to the secret of the creation of plants, and that it is the simplest thing one could imagine. The archetypal plant will be the strangest creation in the world, for which nature herself ought to envy me. With this model and the key to it, one can invent plants endlessly which must be consistent—that, even if they did not exist, could exist—and not some artistic or poetic shadows and appearances, but possessing inner truth and inevitability. The same law can be applied to everything living.[15]

In the same letter he reveals the secret of the metamorphosis of the individual plant:

> It had occurred to me that in the organ of the plant which we ordinarily designate as *leaf* the true Proteus lay hidden, who can conceal and reveal himself in all forms. Forward and backward the plant is always only leaf.[16]

Proteus was the Greek god of the sea who could transform himself into any shape or form but who never revealed himself directly. Like this god, the elusive organ of the leaf reveals itself to the senses only in its many fixed and specialized forms. But it also conceals its inner mobile nature in these outer manifestations. Only an active, imaginative thinking can glimpse it in its formative activity.

Thus we witness the birth of a new human capacity: for the first time human thinking becomes capable of entering, concretely, into the creative processes of organic nature. Goethe's exact, imaginative thinking has developed to the point where it can begin to follow the emergence of the various plant families and species out of an archetype. It can now also participate in the metamorphosis of the leaf as the individual plant unfolds. It has gained a strength and mobility that allow it to transcend the given outer world of space and enter into the dynamic dimension of time. This is the realm of becoming, of rhythmical metamorphosis, of life.

The great inner breakthrough that Goethe achieved in Italy also led to a profound transformation of his poetic style. One of the few possessions that he had taken with him to Italy was his drama "Iphigenia." Like the dramas of his youth, it had been written in prose. In Italy Goethe felt inspired to rewrite it—this time in verse form. In Rome he worked at it regularly, preparing a section before going to sleep in the evening and then, upon awaking next morning, writing it out in verse and speaking it aloud, allowing it to sound in its full, rhythmic musicality.[17] Goethe was no longer satisfied with the raw, prosaic expression of thoughts and emotions. He felt the need to allow the language—and with it the experience he was trying to convey—to breathe with a rhythmical life of its own.

Just as he was now able, in his botanical studies, to enter into the dynamic formative processes inherent in the plant world and see it all as a greater unity, as a dynamic flow of variations on a central, archetypal theme, thus he now also sought the experience of the unifying, carrying, rhythmical flow of the spoken language: the experience of the dynamic, formative life of speech itself.

Goethe's breakthrough can be further characterized by his profound response to the remnants of classical Greek art which he saw in Italy. Of these he wrote:

> These exalted works of art are at the same time the most sublime works of Nature, brought forth by human beings in accordance with true and natural laws; everything arbitrary, fanciful, falls away, here is necessity, here is God.[18]

Great artists, Goethe felt, are literally in touch with the deepest lawfulness, the essences of things:

> Style, however, is based on the deepest foundations of insight, on the being of things—to the extent that it is granted us to perceive this essence in visible and tangible form.[19]

Wherever this essence, this deepest lawfulness, is able to express itself freely and fully, there beauty arises:

> The beautiful is a manifestation of hidden natural laws which would otherwise forever have remained concealed.[20]

The artist has an intuitive grasp of these hidden laws and feels the urge to give them expression:

> When nature begins to reveal to us her open secret, we feel an irresistible longing for her worthiest interpreter—art.[21]

Goethe sees in the given world, in the sense-perceptible, a partial, imperfect revelation of the imperceptible creative forces that have brought it about. He sees great art as a further revelation of these same forces—of the creative forces that are at work in nature, in the human being, and in the world. From now on his poetic work will strive to reveal these forces in an objective artistic way. His science will strive to grasp the lawfulness, the nature, of these forces as they manifest themselves in the world and to formulate the results of this research in such a way that others can follow them step by step.

In trying to assess the significance of Goethe's "rebirth" in Italy we see, on the one hand, when we look at his own development, the two polar aspects of his nature that might be called the masculine and the feminine (as he inherited them from his parents), the objective and the subjective, or the scientific and the artistic, each undergoing a transformation as they approach one another and enter into a kind of marriage. Through years of exact scientific observation and practical experience, Goethe's inborn playful, creative imagination had become a disciplined faculty capable of objective scientific research. He was now able to see in the plant world the playful, creative activity of divine nature. In a letter to Frau von Stein he described his perception of the archetypal plant: "It is a beholding of the essential form with which nature always only plays and, playing, brings forth the manifold forms of life."[22] With his inner breakthrough in Italy, Goethe's imagination sprouted as it were into new, objective activity, following, participating in, and recreating the processes of plant formation so vividly that the formative principles inherent in the plant kingdom stood ever more clearly before his inner eye. Goethe called this new capacity the "perceptive power of judgment" (anschauende Urteilskraft), "exact, sensory imagination" (exakte, sinnliche Phantasie), or simply "perceptive thinking" (anschaundes Denken). At the same time we see in Goethe's poetic work and in his whole conception of art the opening up of sources of more objective inspiration.

In Italy Goethe's scientific work breaks through to imaginative creativity. His artistic work breaks through to a new objectivity. Both aspects of his nature—the great polarities of all human spiritual striving—experience a deepening, an enhancement, as they approach and fructify one another. The fruit of this fertilization is the entry into a new dimension of reality: a realm of creative, formative forces that is both hidden and revealed to the senses through its constant activity in the organic world.

Goethe in Italy, by J. H. W. Tischbein

When we look at Goethe's Italian Journey from an historical perspective, we see these newborn capacities and the new realm of experience that they open up as seeds of a new culture, as the starting point of a complete transformation of our civilization.

Goethe's Italian Journey was in many ways the high point of his life. All the forces of his being seemed to converge and pour forth in their greatest fullness, strength, and freedom. Looking back over his life, he said as an old man:

> Yes, I can say that only in Rome have I felt what a human being actually is.—Never again have I attained this height, this joy of sensation; compared with my state in Rome, I have actually never become happy again.[23]

Weimar Again

AFTER two years in Italy Goethe returned to Weimar in the summer of 1788. The extent of his inner transformation was so great that he returned a stranger as it were and was unable to take up his life as he had left it two years before. Things had hardly changed in Weimar, while Goethe had discovered a whole new world of experience. There was little interest or understanding for the inner treasures that he brought back with him and that he longed to share, test, and further elaborate with others. Herder, unable to renew himself in middle age, had grown fixed in his ideas. Even the friendship with

Frau von Stein had faded. Goethe created something of a scandal in Weimar when he fell in love with Christiane Vulpius, a young, uneducated girl "of the people," and took her into his house. She bore him a son, and they were later married and lived happily together until her death in 1816.

Goethe withdrew from many of his responsibilities at the Court. Most of his time was now spent in serious pursuit of his broad scientific interests. In 1790 he published his "Metamorphosis of Plants," a thorough, detailed presentation of his observations of the dynamic principle of plant formation. In 123 succinct paragraphs Goethe shows how each of the various organs of the plant, from the seed through the blossom to the new seeds, is a transformation of one basic, ever-changing organ that expands and contracts as it metamorphoses.

When we follow the growth of an annual plant, we experience that it is not a continuous but a rhythmical process: When the seed germinates, the sprout throws off its dead husk, sending a root down into the earth as the simple seed leaves unfold upwards toward the sun. After a pause, a new stage begins: a shoot with more characteristic leaves emerges and stretches upwards. At each node there is an expansion into the leaf followed by a contraction in the vertical growth of the next stem segment. As the rhythmical unfolding of vegetative leaves progresses, we observe a greater rhythm: an expansion from the small, rounded seed leaves into the larger, fully elaborated vegetative leaves and then, toward the end of the shoot, a contraction into the tiny leaves of the tightly closed bud. Here, as in the seed, a concentration of the plant takes place. The vital phase of vegetative growth has come to an end; and again there is a pause while, hidden within the bud, the next phase is prepared through a process of intensive transformation. Then, in a new burst of expansion, the bud opens into the blossom with its brightly colored crown of petals. It is in the blossom that the plant reveals its innermost nature. Its beautiful form, its vibrant color and sweet scent have always been admired as the most sublime expression of our deepest feelings. In the blossom the life process of photosynthesis has been repressed, and qualities of a higher world are revealed in an ephemeral display. As in the preceding vegetative stage, an expansion and contraction also takes place in the blossom: the protean leaf-organ expands in the petals and contracts into the stamen. Soon petals and stamen wilt and fall away. After fertilization, growth resumes—but now turned inward in the expansion of the swelling fruit within which the final contraction of the plant takes place in the formation of the seeds. Now the entire plant dies and decays. The seeds, however, within their hard husks, survive as dormant germs of new life until the next spring. Thus the annual plant unfolds in three successive stages of expansion and contraction: the sprouting stage, the blossom, and the fruit. After one stage has come to an end in its contraction, a new impulse enters in, leading to the unfolding of the next stage.

Finding it hard to interest his feminine acquaintances in plant metamorphosis, Goethe resorted to poetry. His poem "The Metamorphosis of Plants" lifts the reader into a vivid experience of the dynamic processes of plant life.

Goethe found even less understanding for his new classical poetic style and for his conception of art as a revelation of a deeper lawfulness. The German public was still reveling in the powerful, raw emotion of Storm and Stress. They were especially enthusiastic about a young poet by the name of Friedrich Schiller, who had outdone all the others with his explosive drama "Die Räuber" (The Robbers) and the enthusiastic call for freedom in his recent play "Don Carlos." Goethe, who had passed this youthful stage in his own development, felt a strong aversion to this raw, impetuous style, which he now saw as subjective and immature. Already at the age of forty, Goethe seemed to have transcended and outlived his time. The years following his return from Italy were overshadowed by spiritual loneliness. There was no one with whom he could share his own deepest strivings.

Friendship with Schiller

FINALLY, in 1794, just before Goethe's forty-fifth birthday, he met Schiller as they were leaving a lecture on science in the university town of Jena where Schiller lived. They exchanged comments on the lecture and were soon engaged in a conversation that became one of the greatest meetings of minds in central European history. Schiller deplored the fragmented manner in which the phenomena had been presented in the lecture. Goethe then spoke of an approach that sees nature as a living whole out of which the individual phenomena emerge. They entered Schiller's house where Goethe launched into a lively presentation of the metamorphosis of plants and finally resorted to an illustration in the form of a characteristic sketch. Schiller, who had followed Goethe's presentation of metamorphosis with tremendous interest and a powerful, penetrating grasp, now shook his head and said, "That is not an experienced perception; that is an idea." Goethe, taken aback by the immense gap that separated their world views, nevertheless collected his wits and responded, "How wonderful that I have ideas without knowing it, and that I can even see them with my own eyes."

Goethe's thinking did not separate itself from his perceptions. His thinking was always concrete perceiving—-it never left the phenomena. It never became abstract. And his observation was always deeply permeated with attentive, participatory thinking. Metamorphosis was for him a process that he observed in the plants. He experienced it as an objective percept just like the percept of the various leaves, petals, stamens, and so on. He did not realize, perhaps until this first meeting with Schiller, that the metamorphosis from leaf to calyx, to petal, to stamen is in fact not an outer percept but a thought process: that the metamorphosis of the plant is in fact an idea. Schiller, on the other hand, did not perhaps realize until this first meeting with Goethe the difference between abstract or speculative ideas and such ideas as plant metamorphosis, which are a spiritual perceiving of objective, natural processes.

Are not, indeed, the deepest ideas arrived at by the human subject at the same time the deepest foundations of the objective world? Exactly one hundred years later, in 1894, this relationship between the two sources of human cognition—subject and object, thinking and perception—was further clarified by Rudolf Steiner in his *Philosophy of Spiritual Activity*.

Schiller, the fiery idealist and herald of human freedom, was a powerful advocate of the free, thinking forces of humanity that awaken us to inner independence and the full dignity of selfhood. Goethe, on the other hand, was the great observer of life who learned through experience and penetrated ever more deeply into the lawfulness of the world. Emerson wrote of Goethe: "The old Eternal Genius who built the world has confided himself more to this man than to any other."[24] If, according to Emerson, Goethe was secretary to the Genius of the universe, then we would have to call Schiller "secretary to the Genius of humanity."

The ensuing friendship between these two great antipodes of human nature became one of the most stimulating, spiritually productive relationships that we know of. Because they were so different, they saw each other clearly and objectively. Shortly after their first meeting Schiller wrote to Goethe:

> Your observant gaze, which rests so peacefully and purely on things, never puts you in danger of being led onto the wrong paths onto which both speculation and arbitrary imagination, which is regulated only by itself, so easily stray.

And later, in the same letter:

> You pull together the whole of nature in order to grasp some particular phenomenon. You seek the

> explanation of the individual in the totality of nature's manifestations.[25]

Schiller observed in Goethe's cognitive approach to the world a holistic scientific approach such as is at last being called for today. And, he saw it in a larger context. He saw in Goethe the rich soul of an ancient Greek who, born into our prosaic, fragmented modern world, is faced with the task of recreating out of himself, through his own thinking activity, a living world view such as that of the ancient Greeks, one in which the divine is experienced in nature and in the here and now. It is significant, in this context, that the origin of our modern word "idea" is the ancient Greek verb "idein," which means "to see." Schiller must have immediately recognized the ancient Greek spirit in Goethe when, in their first conversation, Goethe insisted that he could see his ideas, that they were not abstractions. Schiller understood Goethe's deepest nature and striving, saw its immense significance, realized that he stood at the beginning of an immense and wonderful task, and urgently encouraged him to follow his star.

At last Goethe had found the spiritual companion whom he needed. For him, this friendship meant "a new springtime in which everything germinated happily in every sphere and burst forth from sprouting seeds and shoots."[26]

Schiller encouraged Goethe to pursue his scientific research and to renew his work on his great Faust drama and on his novel Wilhelm Meister. Much of Goethe's greatest, ripest work might never have come about without Schiller's inner support. Goethe, on the other hand, helped Schiller plan and execute, among other works, his great dramas on the genius of the various European nations. Through Schiller, Goethe once more gained contact with the German public, and together these two spiritual giants became the undisputed cultural leaders of Europe.

The extraordinarily productive friendship between Goethe and Schiller marked the high point of modern German culture. Goethe's spiritual realism and Schiller's fiery idealism radiated out from central Europe as beacons of cosmopolitan humanity. The archetypal image of humanity engendered by their friendship is best expressed in Schiller's philosophical work, "Letters on the Aesthetic Education of Humanity" (1794-5) and in Goethe's "Fairy Tale" (1795), which speaks in purely imaginative, symbolic pictures. With a spiritual insight more encompassing and deeper than any offered by our popular culture today, these works, each in its own way, portray the human condition poised between the world of the senses and the world of the spirit, and they show how, through free, creative spiritual activity and self-transformation, the human soul can bring forth a middle realm of experience in which the two originally disparate worlds are harmoniously united.

By 1800 the strain of wide-ranging cultural activities began to show itself, and Goethe's health gradually deteriorated. In the spring of 1805 both Schiller and Goethe fell seriously ill. Schiller, whose fiery spirit had completely burned out his physical organism, died on May 9th. Goethe was devastated by the news. He shut himself up for days. For weeks he hardly spoke a word and did not touch a pen.

Goethe's illness in Leipzig had brought him close to physical death. In Weimar, before he fled to Italy, he had experienced what can be characterized as an inner soul death. When Schiller passed away Goethe again went through a death experience of his own. Again he had been very ill; the soul devastation must also have been overwhelming: from then on the mention of Schiller's name brought tears to Goethe's eyes. And yet there was an even greater aspect: Goethe had not only lost a dear friend, he had lost the only human being who truly understood him, who had raised him out of resignation into an unhoped-for period of productivity, whose nature deeply complemented his own, and perhaps most importantly, together with whom he might have built lasting foundations for a central European culture that could have withstood the onslaught of materialism, industrialism, and nationalism that Goethe already foresaw. It is possible that, at Schiller's death, Goethe had a premonition of the death of the deeply human, yet objective, cosmopolitan German culture that he and Schiller had endeavored to foster, a culture seeking consciously to perceive the concrete working of the creative spirit in all of nature and in the human being. Instead, German culture in the nineteenth century split into the sweet otherworldliness of romanticism on the one hand and the harsh "thisworldliness" of realism and naturalism on the other, foreshadowing a split in Western culture that has grown deeper in the twentieth century and that is crying out to be healed today. The tragedy of the German people is reflected in the fact that the powerful spiritual impulse that lived in its great cultural leaders around the year 1800 had faded into a pale memory by the year 1871 when Bismarck united Germany. Instead of through true humanity and insight into its spiritual tasks, Germany was united through "blood and iron."

A New Science of Color

AFTER the crisis around Schiller's death was over, Goethe recovered, and, miraculously, new forces of regeneration welled up within him. During the first five years of this period of renewal, we find Goethe working with increasing intensity on his "Farbenlehre," his study of color, which he published in 1810. This, his most comprehensive scientific work, had begun shortly after his return from Italy in 1789, and with few interruptions, it continued to occupy him for the next twenty-one years, and then intermittently, until the end of his life.

Goethe's views were based solely on thorough observation. One of his central principles was: "Let us not seek for anything behind the phenomena; they themselves are the theory."[27] Thus, in the second chapter of his "Farbenlehre," he describes the genesis of colors in the atmosphere:

At noon, the sun appears, colorless in its brightness. This changes as the sun begins to set: at first it turns yellowish, then gold, then finally orange, and even deep red, as it sets on the horizon. As the sun sets and its light has to penetrate an ever-thicker section of the earth's hazy atmosphere, the "warm" side of the spectrum appears.

Later, in the sixth and final chapter, Goethe describes the qualitative or inner, aesthetic-moral experience of color sensation (die sinnlich-sittliche Wirkung der Farbe), which can only be indicated here: Through the slight resistance of a translucent medium, pure light is softened to a cheerful, sunny yellow, which has a light, expansive quality. As the light has to overcome more and more resistance, its color gains in warmth, strength, and intensity, growing grander and more powerful as it passes from gold to fiery orange. Here the color reaches its greatest energy and can take on an almost violent character. The deep, ruby red, finally, is the greatest achievement of the light as it were; a new, powerful serenity is attained as the greatest intensity overcomes the greatest resistance. This noble color seems to glow with an intensity purified and deepened by suffering. It is, perhaps, the most deeply satisfying and sublime of all the colors.

The "cold" side of the spectrum appears when we view the darkness of outer space through the turbidity of the illuminated atmosphere. Thus, on a clear day, the sky overhead appears dark blue; closer to the horizon it lightens, sometimes to light turquoise, and finally it appears white where the hazy atmosphere becomes impenetrable to our gaze. Rising to an altitude of twelve miles, we see the sky above us turn an ever-darker blue, until it finally appears a deep violet. Above eighteen miles, the sky is black, even though the sun is shining.[28] Just as Goethe characterizes the "warm" colors as active, expansive, "plus" in energy, and musically, as major in character, he describes the "cold" colors as evoking a passive, minor mood of longing, as "minus" in energy. Thus blue can give rise to a sad, wistful feeling, drawing us out into itself, into the distance. We experience an intensification in the violet: the indefinite longing of the blue becomes a mood of fervent, prayerful devotion. Thus Goethe experienced the atmospheric colors in an aesthetic-moral way as "the deeds and sufferings of the light."

In his study of color Goethe was able to show how the polarities of light and darkness, when viewed through various translucent media—air, smoke, milky glass, and so on—give rise, under the right conditions, to the "archetypal phenomena": the warm and cold sides of the spectrum. In the realm of chemistry, the acids show an affinity to the warm side of the spectrum whereas the bases are related to the cold side. Particularly beautiful are the experiments with prisms in which, through the overlapping of the two spectra, the yellow and blue produce the passive green, while their intensifications, red and violet, produce a pure peach-blossom red. In this way the ends of the two spectra connect, giving rise to a color circle. Through a process of intensification, the yellow and blue raise themselves up as it were to meet in the beautiful pure red; through passive mixture they form green. Green and pure red are complementary colors, as are all the opposing colors in this color circle. Thus, throughout the world of color we experience the dynamic interplay of polarities.

Only a superficial indication of this vast and fascinating work can be given here. Suffice it to say that it leads to an amazingly diverse yet unified view of the world of color. And, as color speaks not only to our senses but to our feelings, the ordered world of color can be experienced as a language of the soul as well as of the sense perceptible world. For the color experience itself is a bridge between the outer, objective world and our inner, subjective being. Light—so much like our own spirit—is invisible. Like the latter, it reveals its inner nature only through its "deeds and sufferings": through the vibrant language of color.

How do we experience the aesthetic-moral qualities of the colors? Are they not an entirely subjective response to an outer stimulus? Goethe's experience was that feelings, when cultivated, can become transformed into organs of highly differentiated perception. Just as the rudimentary organ of light sensation that we find in the lower animals has developed into the highly evolved optical instrument of the human eye, thus the sensitivity of the human soul is capable of evolving from a primitive sense of our own pleasure and pain into an organ that is responsive to the whole spectrum of qualities displayed by the manifold beings in the world around us. Thus the elemental power of our passions—desire, fear, sympathy, and revulsion—can gradually be transformed into compassion: into the selfless capacity to feel-with, to experience and perceive the inner nature of the other. This transformation of our feelings into transparent organs of perception takes place as we are able to permeate them with ever greater powers of consciousness. Our feelings then no longer rule over us as half-conscious moods. They become balanced organs of active perception.

By participating in the specific quality of a color, a sound, a plant or animal, we bring it to life within ourselves. This vivid inner experience is then not subjective: for through it we come to perceive the essential nature of the object in question. Goethe wrote: "Every new object, well observed, opens up a new organ of perception within us."[29] Only through what reverberates and comes to life within us can we perceive the inner, essential nature of the object. The object presents itself to our senses as a riddle: it gives us all the clues. However, we can only find the solution within ourselves. The solution is nothing other than the essential being of which the clues are the manifestations. This essential nature of the object becomes inwardly tangible in the form of a feeling that is so permeated with consciousness, so clear and distinct, that it becomes a cognitive feeling. At this stage our feelings become knowledge or insight; our insight is permeated with feeling. We now have the deeply satisfying experience of seeing the object, in all it outer manifestations, as the revelation of its living inner lawfulness. It is as though we were beholding the object for the first time. Now, at last, we are able to see it in its full, dynamic reality. Just staring at the color yellow, for example, will reveal nothing of its character. Only through our inner participation in its quality will we come to experience its essential aesthetic-moral nature. Thus we are only able to perceive the essential character of an object when it awakens as an inner experience within us. Goethe drew attention to this principle in his *History of the Study of Colors* through a little verse that goes back to the German mystic Jacob Böhme:

Sunlike itself the eye must be,
else it no glint of sun could ever see.
If, in us, there dwelt not God's own might,
in things divine we could take no delight.[30]

Within each human being, then, as within a microcosm, lie dormant all the essential forces that have shaped the universe. In the macrocosm these forces manifest themselves, partially, as sense experience. Sense experience is the catalyst, the clue, for the actively sensing human being to awaken the experience of the corresponding universal forces within him- or herself. The forces that awaken in us are the organs through which we perceive the essential forces in the world. Thus the human soul evolves towards an ever richer, more perceptive inner life by directing its attention outwards and then inwards, and then outwards again in a kind of rhythmical breathing activity.

With his study of color Goethe laid down the foundations of a qualitative science. Analytical, quantitative science deals with the purely physical, spatial aspect of the world. The science of metamorphosis explores the formative principles of the organic world that live in the dimension of time. Goethe's science of color is research into yet deeper dimensions of reality that manifest as specific qualities.

Engaging in science on this level is equivalent to learning nature's language. We are already conversant with the "body language" of the human being: with the physiognomic language of a smile or a scowl, of a gesture or a particular posture. We understand it so readily because we ourselves are constantly using it, consciously or not, to express our own inner state of being. In order to experience the language of nature, however, we must develop the capacity to live into its qualities as intimately and strongly as we are able to live into the physiognomy or gesture of another human being. As a poet, Goethe possessed the perceptive feeling capacity to experience this language and to bring it to expression in his writing. However, he also felt the need to order his experience, and to grasp its inner lawfulness with his thinking. Goethe was therefore predisposed to develop a qualitative, or physiognomic science.

Through his studies, Goethe achieved an understanding of color as the deeds and sufferings—the vibrant language—of light. He planned to undertake a similar study of tone. Eventually this scientific approach will advance toward an experiential understanding of all things as the physiognomic expression of their essential inner being, whether they be minerals, trees, animals, people, landscapes, or planets. Then our present alienation from nature will be overcome, and we will be able to see through the appearances into the heart of things.

Of all of Goethe's works, his study of color has been and remains the most controversial. For, from the outset, Goethe's approach to the world of color stood in direct opposition to the abstract, reductionist approach of modern physical science as it was at that time represented by Isaac Newton.

Just as reductionist biology seeks to explain the plant atomistically through the mechanisms of its genetic structure and biochemistry, thus Newton and his successors have sought to explain the phenomena of light and color as the result of mechanical or electromagnetic forces. Goethe, on the other hand, came to see the dynamic world of light and color as resulting from the manifold working of primal, irreducible archetypal phenomena in the same way that he had come to see individual plants as continuously metamorphosing wholes and the plant kingdom as the wonderfully varied manifestation of the archetypal plant. In explaining color, Newton literally lost sight of the colors themselves. Goethe, however, entered more deeply into the colors and learned their own language. Goethe's phenomenological, holistic approach penetrates to the lawfulness of the phenomena themselves, to their deeper, spiritual dimensions. Newton's reductionist approach, on the other hand, reduces the phenomena to their material substratum: to molecules, waves, or particles and then proceeds to explain the former in terms of the mechanical interaction of the latter.

These two views represent two perspectives that need not be mutually exclusive, however. Goethe's view of color excludes the presence of corresponding electromagnetic phenomena as little as the study of sound excludes the presence of airwaves, or as dynamic morphology excludes the biochemical activity in the organism. Goethe and Newton were focusing on different realms of reality. Whereas Goethe's approach comes to see the physical as the footprint of the creative spirit, as a necessary condition for its appearance, contemporary mainstream science, following Newton's lead, generally declares the material aspect to be the only reality. Reducing nature's phenomena to their material substratum, however, though necessary for technical understanding and use, does not do justice to the actual phenomena themselves. On the contrary, it in effect abolishes them and fails to see their essential nature. Put into practice, radical reductionism ultimately leads to the degradation of life.

In the sharp controversy ignited by these radically opposed positions, Goethe felt that he was representing Nature herself against a view that reduced her to a meaningless mechanism and failed to perceive her phenomena as the vibrant revelations of her being. Goethe's feelings were deeply engaged in his science of color. Though he was scrupulous in his methodology, his relationship to the phenomena might well be described as religious. This was reason enough for many cool, detached reductionists to dismiss his views out of hand. What they failed to realize is that Goethe was allowing the phenomena to speak their own language, that their language can only be experienced by perceptive feelings, and that depth of perception can only be attained through a corresponding depth of feeling. This explains the passionate vehemence and tremendous, patient expenditure of energy with which Goethe put forward and defended his views. His own assessment of the importance of his study of color speaks clearly for itself in this conversation reported by Eckermann in 1829:

> "As for what I have done as a poet," he would repeatedly say to me, "I take no pride whatever in it. Among my contemporaries there are excellent poets; poets more excellent have lived before me, and others will come after me. But that in my century I am the only person who knows the truth in the difficult science of colors—of that, I say, I am not a little proud, and here I have a consciousness of superiority to many."[31]

Through his work on the metamorphosis of organisms and his qualitative study of color Goethe introduced two substantially new levels of scientific research. That these two new approaches have been misunderstood, rejected, and ignored until today is due in part to the fact that the reductionist approach, which has its validity in its own realm, is seen as the only valid scientific method. The deeper levels of reality with which Goethe was concerned—the levels at which the phenomena manifest their own life and being—are, if not rejected outright, considered to lie outside the domain of science. That these two new levels of science have not been practiced more widely is also due to the fact that their practice requires new, transformed human capacities of perception: enlivened, participatory thinking, and highly perceptive, objectified feeling.

The relationship between these three levels of science can be illustrated in the analogy of "the book of nature." Here the phenomena of nature are compared to the printed words in a book. On the first level, reductionist science analyzes the letters and seeks to explain their form in terms of the molecular structure of the ink. It has become very adept at the technological manipulation of ink and paper. On the second level, the forms of the letters are soon perceived as manifestations of higher principles. They are observed, compared, and ordered. The relationship of one letter to the next within the context of a whole word is established through a kind of "sounding out" process. Words are seen as parts of sentences; sentences are seen in the context of paragraphs. Thus we come to grasp the dynamic formative principles, the grammar, of the texts. Only on the third level, however, does the actual reading begin. For here the words become transparent and their actual meaning is revealed. We find ourselves engaged on a deeper level as we begin to experience the story-content of the text. This new, inner dimension of the book comes to life not only in our thought but also within our qualitative life of feeling.

As any teacher knows, reading is not a passive affair. It is an act of the imagination, and it requires considerable inner activity and inner engagement on the part of the reader. The detached onlooker will see only meaningless letters. The more deeply and rigorously we engage ourselves in the reading activity, the greater our understanding of the printed text will be. In 1975, a year before his death, the chemist and philosopher Michael Polanyi wrote of science: "Its method is not that of *detachment* but rather that of *involvement.*"[32] Goethe developed methods of objective involvement. Lest he fall into a subjective interpretation of nature, he was always wary of reading anything into her phenomena. "In observing nature on a scale large or small, I have always asked: Who speaks here, the object or you?"[33]

Seen in the context of this analogy, our modern, reductionist science is illiterate. It can only exploit the material aspect of nature. But doing so without an understanding of nature's own deeper lawfulness, it must ultimately become arbitrary and destructive. Only on the basis of an insightful "reading" of nature's "script" will a healing of our civilization be possible. Goethe's morphology and his study of color were the first scientific attempts to "read" nature's "script" on the levels indicated here.

Old Age—New Beginnings

AFTER the publication of his study of color, this productive phase of Goethe's life culminated in his relationship with Marianne von Willemer, and in an unprecedented outpouring of poetry, collected in the twelve books of his West-östlicher Divan. After 1815, however, his creative forces seem to flag, and in the next eight years he resigns himself to iron discipline, an almost pedantic carrying out of his various duties and a frigid, almost icy relationship with all but a small circle of close friends.

In 1823, in his seventy-fourth year, a new crisis shakes his life to its foundations: in February he becomes suddenly ill and is very close to death. He recovers and that summer goes to Carlsbad as is now his custom, feeling better than he has for years. It is here that he falls deeply in love with Ulrike von Levetzow, a charming girl of nineteen, young enough to be his granddaughter. The ice of old age thaws, a new youth, a new rapturous springtime breaks forth from within! Goethe finally asks for her hand; there is hesitation, probably on the part of the girl's mother, and Goethe now realizes that, once again, his heart has placed him in opposition to the realities of life circumstances, and he breaks off the relationship forever. It is for him as though he were irrevocably renouncing life with all its fullness, joy, and wonder. That winter he is again very sick. Is it the end?

Death and illness yield once more to the gentle springs of new life. Renunciation and the poetic fruit of his experience, the beautiful "Marienbader Elegie," have cleansed his soul and raised it—opened it—to a new level of transparency, granting him a clearer, deeper look into life's "open secrets." His last eight years are remarkably productive. Outwardly, his life is restricted to his home in Weimar; but spiritually he reaches out further than ever before. He writes from three to four hundred letters a year, the conversations recorded by Eckermann fill a whole volume, profound lyric poetry arises, he finishes his novel *Wilhelm Meisters Wanderjahre* (The Journeyman's Years). His greatest achievement, however, is the completion of the second part of his Faust drama, the great work that he began in Strassbourg when he was twenty-one and which accompanied him throughout his life. From the outset, the whole inner conception of this great drama of human striving had lived within him as a seed.

Throughout his life, it had unfolded, part by part. But only in his last years could the ripening process take place that was necessary for its completion. It is as though another veil falls before his eyes, allowing him insights into a yet deeper dimension of reality. In 1830, two years before his death, the eighty-year-old Goethe wrote to his friend Zelter:

> Perhaps I may whisper it in your ear: I am experiencing the happiness of having ideas unfold in my old age, which to develop and bring to realization might well be worth living life over again.[34]

We must bear in mind that, for Goethe, ideas were not abstractions but real principles: "That which always comes to manifestation and therefore reveals itself to us as the lawfulness in all manifestations."[35] One of the ideas with which Goethe concerned himself increasingly in his last years was the idea of the "entelechy" (a Greek word meaning "that which carries its end or goal within itself"). Goethe had always seen the concrete unfolding of an organism as the result of the interaction of an inner, self-active entelechy with the outer conditions of the environment. Only in the human being, however, can an entelechy actually come to full consciousness.

The human entelechy is a purely spiritual source of creative activity. Although it is constantly revealing itself through its activity, it can never reveal itself entirely, for it lives as a deeper, future-bearing impulse of self-transformation that can only express itself to a limited degree through the vehicle of our psychological and physical nature. But it is the entelechy, or "I," that gives us our inalienable human dignity and which is the basis for the free spiritual community of all humanity, for it transcends race, nationality, and sex.

The realm of the entelechy is the realm of the idea as quoted above—of the idea as creative power. It is the realm of the archetypal plant, of light, of the human "I." It is that which manifests itself; and it therefore also exists independently of its manifestations as purely spiritual being. As such it transcends the phenomenal world in which Goethe, as a down-to-earth empiricist, was at home. Nevertheless, Goethe did explore this purely spiritual realm insofar as it presented itself directly to his own inner experience and insofar as he was able to perceive its manifestations in nature.

It was the experience of his own entelechy as a wellspring of creative activity that gave Goethe the assurance of his immortality:

> The conviction that we are destined to endure is rooted only in the concept of activity; for if I continue ceaselessly active to the end of my days, Nature is under obligation to assign to me another form of existence, when the present one is no longer able to contain my spirit.[36]

How did Goethe approach this deeper dimension of reality in his old age? And how does it relate to the dimensions he had already begun to explore?

Goethe experienced his immortal entelechy as will in its highest form: as creative impulse. Will is unconscious, creative spirit: the central driving force in our deeds and in world events. Feeling is a semiconscious, qualitative experience of our own state and of the world as revelation of the spirit. Thinking rises still further up out of the unconscious sea of world will activity; it separates itself out as an organ of full consciousness, capable of reflecting or picturing the world. But, initially, our reflective thinking is divorced from the deeper sources of reality from which it has freed itself. Our abstract concepts are dead—devoid of the life and vitality that rules the world.

This is the situation in which modern humanity finds itself today. This is the situation in which Goethe's Faust, the representative of modern humanity, finds himself at the beginning of the drama. Writing in Strassbourg, the 21-year-old Goethe passionately portrays Faust's despair over the shallow impotence of the intellect. He formulates Faust's deepest striving: to "see the secret force" that "hides in the world and rules its course." However, when Faust impetuously conjures up the fiery Earth Spirit, he is overwhelmed and repulsed by its power. He is unable to withstand the conscious experience of the creative spirit. Nevertheless, our fully awake thinking, isolated from the powerful forces of the world, thin and weak as it may seem, must be the starting point of our cognitive journey back to the sources of reality.

When Goethe was experiencing his profound renewal in Rome, he wrote the "Witch's Kitchen," the scene in which

Faust is rejuvenated. Here Faust finds himself amid the dark, demonic world of the nordic witches. It is a chaotic, fragmented world, which Faust abhors. However, he is drawn to something that seems to come from a different realm:

> FAUST *[who has all this time been standing before a mirror, now going near it, now going away from it]:*
>
> What do I see? What form divinely fair
> Within this magic mirror is revealed?
> Oh lend me, Love, thy swiftest wing and bear
> Me hence into her wondrous field!
> Alas! If from this spot I dare
> But stir, or if I venture to go near,
> Then dim as through a mist doth she appear!
> The fairest image of a woman! Can it be,
> Is it possible? Can woman be so fair?
> Must I in that recumbent body there
> Behold of all the heavens the epitome?
> Can one so fair be found on earth?[37]

Is this not the quintessence of Goethe's experience in Rome, which he contrasts here with the world of the nordic witches? It is the picture of Helen of Troy, the epitome of Greek beauty, the living spirit of Greece, which Faust sees in the mirror. In Italy Goethe experienced the world of archetypal formative forces—both in the plant kingdom and in the remnants of Greek art. And he experienced it with the living forces of a transformed thinking—as in a mirror.

During his last years Goethe brought Faust closer still to Helen. At the emperor's palace Faust conjures up the scene of the abduction of Helen by Paris with which he intends to entertain the court. In order to find Helen, he must enter the realm of the "Mothers": the realm of creative spiritual archetypes. This time, when she appears, her movements sound as music.

FAUST:

Have I still eyes? Is Beauty's spring, outpouring,
Revealed most richly to my inmost soul?
My dread path brought me to this loftiest goal!
Void was the world and barred to my exploring!
What is it now since this my priesthood's hour?
Worth wishing for, firm-based, a lasting dower!
Vanish from me my every vital power
If I forsake thee, treacherous to my duty!
The lovely form that once my fancy captured,
That in the magic glass enraptured,
Was but a foam-born phantom of such beauty!—
To thee alone I render up with gladness
The very essence of my passion,
Fancy, desire, love, worship, madness![38]

This second meeting clearly takes place in the realm of heightened, perceptive feeling. But Faust has not yet fully transformed his powerful emotions—desire drives him to intervene in the action, whereupon there is an explosion, leaving Faust unconscious as the scene dissolves.

Faust would never have been able to show to others what he beheld in the mirror. For it was the harmonious weaving of formative forces that he beheld in his own thinking. But now, when Faust experiences Helen through his feelings, he is able to reveal her being in all its beauty to the courtiers as in a kind of theatrical display. Here her spiritual essence appears as revelation. This is a powerful experience, but it is not yet the experience of the reality of her being.

This is finally achieved in the culminating Third Act into which Helen enters, no longer as a condensation of nebulous mists but in full, living reality. Her poetic exchange with Faust leads to their union; and she bears him a son, Euphorion. This is a symbolic depiction of spiritual communion on the level of the transformed, fully conscious will. Faust has raised the power of this sleeping capacity into the cognitive realm, transforming it into an organ of awake perception. Only our will, the central force of our being, can perceive the eternal entelechy, the source of creative activity, in the other. Perception on this level, however, fully overcomes the separation of subject and object: it is conscious union of being.

Thus, through Faust's three encounters with Helen, Goethe gives symbolic form to the three fundamental approaches to the divine in nature,[39] which distinguish the three significant stages of his own Faustian striving. The third stage, to which he raised himself in the last years of his life, is characterized by the immediacy of spiritual experience that we sense in the words:

HELEN:

I feel so far away and yet so near
And all too glad I say: Here am I! Here!

FAUST:

I scarcely breathe, words tremble, check their pace;
It is a dream, vanished are time and place.

HELEN:

I feel I'm lived-out sheer, and yet so new,
Blent with thee here, to thee, the unknown, true.[40]

What is it that leads the eternal spirit back through the realm of time into the physical world of space? It is the power of rhythm. Thus the eternal entelechy weaves in breathing, ever metamorphosing movement in and out of the physical world. Perpetual life, bound by the limitations of time and space, would lead to the incarceration of the spirit: to living death. Dynamic, invigorating change and transformation—the creative activity of the entelechy—would become impossible. Already in his youth Goethe realized that "death is nature's device to create more life,[41] that death is a necessary part of the life cycle.

For Goethe, the idea of reincarnation was not only inherent in his larger view of life as rhythmical metamorphosis—it was also confirmed by his own intuitive life experience.[42] The purely spiritual existence of the soul after death Goethe thought of as a time of rejuvenation:

> What a good thing it is that mortals die, if only to shake off the stamp of mortality and come back refreshed as by a bath![43]

Goethe at the age of 77, by L. Sebbers

What is it that refreshes the entelechy and gives it renewed strength for its next incarnation?

During the day we are consciously focused on our outer earthly tasks. At night, however, the earth is veiled in darkness, the stars appear, and our entelechy is drawn out into their peripheral realm—into the unconscious world of sleep: the "inner" world of the spirit, of will. Except for these regular exhalations of sleep, our lives are spent in earthly, outward-directed activity. At death, however, there is a far deeper outbreath: we sever our connection with our earthly body entirely, and our entelechy moves out into the cosmos, where it experiences the rejuvenating forces of the spiritual world. The entelechy then enters its next life with renewed spiritual impulses. Cautiously, Goethe speaks of the entelechy's relationship to the heavenly world in his novel *William Meister's Journeyman's Years* where, in describing the "beautiful soul" of Makarie, he expresses the hope that "such an entelechy not leave our solar system entirely, but that it, having reached its limits, will long to return in order to again be active in earthly life for the benefit of our great-grandchildren."[44]

During the last eight years of his life Goethe was able, by drawing on intuitive powers of perceptive will, to bring his great life's work to completion. While civilization rushed on into the 19th century without him, Goethe passed away peacefully on March 22, 1832.

When, toward the end of his life, a visitor flattered Goethe, attesting to the admiration of all the world for the astonishing productivity of his genius, Goethe turned his large, penetrating eyes upon him and replied: "I know, I know; I also know that I appear to the world as one who has built a ship atop a high mountain, thousands of miles distant from the ocean; but the waters will rise, and my ship will float and sail."[45]

The Underlying Rhythm in Goethe's Life

SOME essential aspects of Goethe's immensely significant life and work now lie before us like the picture of a mountain range. We may now turn to the question as to how these great masses were heaved up above the plain of average humanity. Goethe himself gives us a key to the understanding of his own life in a conversation with Eckermann in which he speaks of productive individuals in general:

> Every Entelechy is a piece of eternity, and the few years during which it is bound to the earthly body do not make it old. If this Entelechy is of a trivial kind, it will exercise but little sway; the body will predominate, and when this grows old the Entelechy will not hold and restrain it. But if the Entelechy is of a powerful kind, as with all men of natural genius, then it will, with its animating penetration of the body, not only strengthen and ennoble the organization but

> also endeavor with its spiritual superiority to confer the privilege of perpetual youth. Thence it comes that in men of superior endowments, even during their old age, we constantly perceive fresh epochs of singular productiveness; they seem constantly to grow young again for a time, and that is what I call a repeated puberty.[46]

The repeated puberties in Goethe's own life and the new capacities that they opened up have been a main focus of this essay. They may, indeed, be considered to be a kind of archetypal phenomenon within the context of Goethe's biography.

When we apply Goethe's scientific method to his own biography, actively participating in its dynamic unfolding and then drawing back to sense the inner quality of its various phases, we gradually become aware of its remarkable lawfulness:

Goethe's birth was, in fact, an overcoming of death. His childhood in Frankfurt allowed for a strong and harmonious unfolding of his gifts. In the sacrifice at his father's music stand we see the deeper impulse of his individuality coming to expression. Whereas his childhood can be experienced as an expansive unfolding, his student years in Leipzig would have to be characterized as an all-out adolescent attempt to plunge into and take hold of the world. But life in "Little Paris" proved to be a dead-end for Goethe. The resulting crisis led him very close to death and home again to a time of withdrawal from life in which a turning inward to the deeper, creative sources of his being could take place.

The youthful Strassbourg years see Goethe awakening to the powerful impulses and forces of his genius, which led to the productivity of his Storm and Stress period. But then, in Weimar, as in Leipzig, Goethe turns to the world. This time he enters deeply into the "objective" realm of administration and of scientific pursuits. Again, however, this finally leads to a death experience.

A momentous rebirth occurs in Italy during a period of escape from the pressures of life in Weimar. The new forces springing from this profound transformation allow Goethe to take hold of his thinking in such a way that it is able to grasp the dynamic principle of metamorphosis and experience the realm of living formative forces. This new phase of his life does not fully unfold until his friendship with Schiller begins in 1794. The last years of their friendship, however, again a period of intense worldly engagement, show the strain of their tremendous exertion, which leads to their illness and to Schiller's death.

Once more, Goethe recovers, and a new period of productivity begins in which he resumes his research into the world of color. He is now able to bring to preliminary completion this pioneering work, which approaches the phenomena as revelation. This mode of knowing requires the perceptivity of a liberated capacity of sensation and feeling that is no longer bound by subjective self-experience—likes or dislikes, pleasure or pain—but becomes vivid revelation of the other. This heightened feeling-capacity was also the source of inspiration that flowed so abundantly into his *West-östlicher Divan*. But then come wintry years of iron discipline and punctilious attention to the demands of the outer world. At seventy-four ravaging illnesses frame Goethe's love for Ulrike von Levetzow, which erupts from depths of immortal youth.

This time, however, renunciation leads to a spiritual springtime, a time of grace in which Goethe is able to attain his deepest wisdom and bring his life's work to completion. In the ripe works of his old age Goethe is able to approach spiritual aspects of the world that can only be grasped when the sleeping will, our deepest and most spiritual capacity, awakens as an organ of perception.

Within Goethe's biography we can discern four dynamic life cycles followed by the first phase of a fifth. Each cycle begins with a "rebirth." New creative forces then pour forth from within, initiating a period of productivity. In this sense, this expansive phase can be called "subjective." When this stage reaches its height, however, there is a turning point and a kind of down-turn into an "objective" phase—a thorough grappling with the world, which requires inner discipline and learning, and is accompanied by a diminishment of inner forces that finally becomes stifling and leads to a death experience. Each cycle can be perceived as a liberating, expansive outbreath followed by a restrictive, often painful contraction, or as a spring rising into the fullness of summer and then falling into autumn and winter.

The following short verse from the *West-östlicher Divan* expresses this breathing experience that Goethe knew so well:

In breathing we two blessings know:
Drawing air in, and letting it go;
One winds us tight, the other relieves;
How wonderfully breathing in all life weaves!
So thank we God when we are pressed
And thank him again when he grants us rest.[47]

The recurring crises in Goethe's life can be compared to the instant between the inbreath and the outbreath. It is a turning point: a moment in which the old can die and a new impulse can take hold. It is a time of metamorphosis.

Nowhere in nature is the principle of metamorphosis more dramatically revealed than in the transformation of the caterpillar into the butterfly: its voracious life as larva having come to an end, the outer skin of the caterpillar hardens into a tough, protective shell, the chrysalis. Within this shell, fully separated from the outer world, and relieved of all outer functions, the transformation takes place. Now, as within the egg, all forces are directed inwards. Finally the chrysalis splits apart and a newly formed butterfly emerges, its colored wings unfolding and expanding as they fill with life-giving juices. And now, "the creature that has lived a wormlike life for weeks and months, or which has apparently been sleeping the sleep of death in its cerements, soars aloft in the air, the companion of the sunlight and the breezes."[48]

This astounding physical occurrence can be seen as an outer symbol of the kind of inner metamorphosis that Goethe went through during his "repeated puberties."

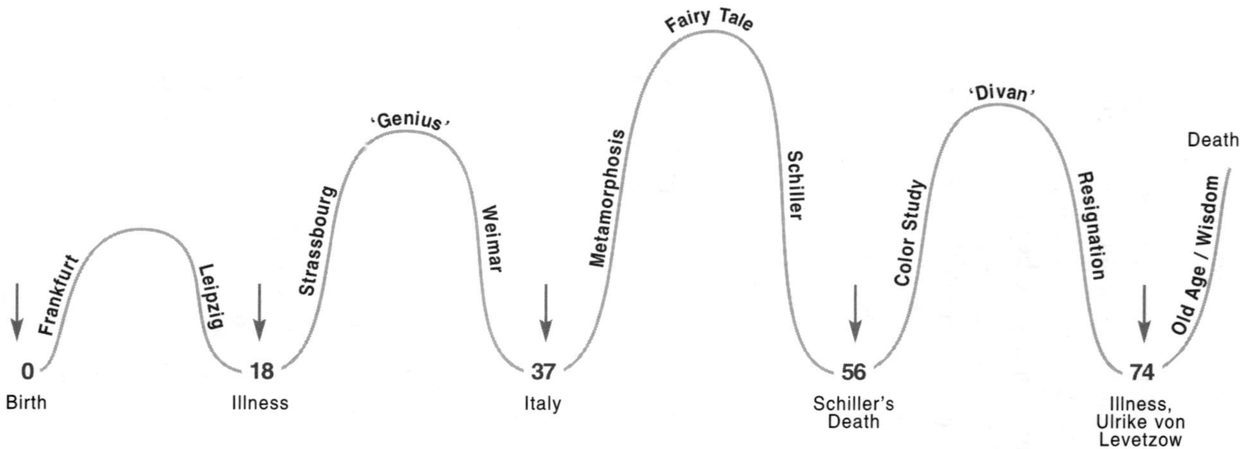

D AGRAM 1   The 18.6-year rhythm in Goethe's life

Diagram 1 indicates the rhythmical unfolding of Goethe's life and is no more than an abstract, schematic map with a very few signposts. The reader must recreate the actual journey—the life experience the dynamic pattern of which it attempts to illustrate.

When we look at the incisive times of "death and rebirth" that mark the end and the beginning of each cycle, we are struck by the fact that they come at regular intervals of a little more than 18 1/2 years.

The regularity and consistency of this cyclical rhythm are remarkable. Is this pattern a coincidence—a rhythm peculiar to Goethe's life alone? Or does it arise out of a greater lawfulness? The soul-breathing process of sleeping and waking is embedded in the greater cosmic-terrestrial breathing process of night and day. Does the rhythmical pattern in Goethe's life correspond to some more subtle breathing rhythm of nature?

The greatest natural rhythms in the life of the earth are those in its relationship to the sun and to the moon. Both were exerting their strongest influence at Goethe's birth: the sun stood at its height, and the full moon stood in opposition. We all know the monthly rhythm of the waxing and waning moon, which governs the tides and is intimately related to vegetative growth and to the reproductive rhythm in the female organism. There is, however, also a greater rhythm of the moon, which requires considerable effort to imagine.

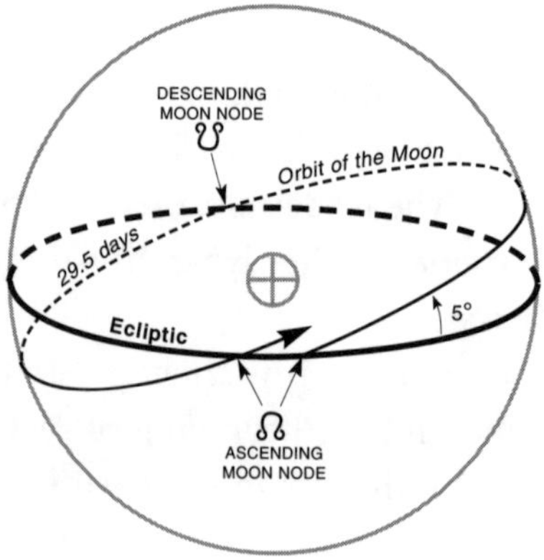

DIAGRAM 2
In a period of about 18.6 years the moon nodes will have completed their revolutions around the ecliptic.

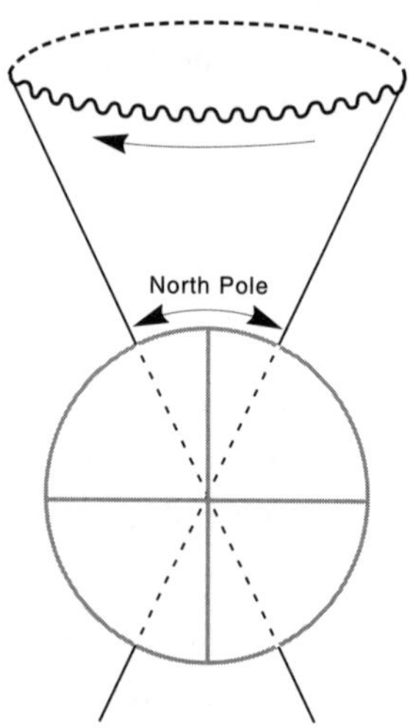

DIAGRAM 3
As the extended earth's axis describes a full circle in roughly 25,770 years, it moves in a wavelike fashion, each wave taking about 18.6 years. Approximately 1,385 waves are completed in each cycle.

As we are dealing with cosmic influences upon the earth, we shall describe the astronomical phenomena as they are seen from the earth. The 29.5-day rhythm already mentioned takes place as the moon passes close to the sun at new moon and then each day falls further behind until it stands opposite the sun at full moon. Then it begins to wane until the sun once again overtakes it at new moon. Each year this dynamic interplay of sun and moon repeats itself thirteen times as the moon moves through the zodiac each time, while the sun moves through it only once. The sun appears to move through the zodiac on a line called the ecliptic. If the orbit of the moon lay on the same plane as the ecliptic, then the moon would pass directly in front of the sun at each new moon, and there would be about thirteen solar eclipses each year. However, the orbit of the moon lies on a plane tilted 5° away from the plane of the ecliptic. Therefore a solar eclipse can only occur when the two planes (and the two orbits) intersect. These two points where the orbits intersect are called the moon nodes (see diagram 2). Now, the orbit of the moon does not maintain its position relative to the ecliptic but gradually shifts around it so that the moon nodes move along the ecliptic, gradually passing through the whole zodiac. The moon completes this cycle, the moon nodes returning to their original position on the ecliptic, in a period of about 18.6 years. In this time the moon's orbit moves through every possible position relative to the zodiac and to the earth.

This greater rhythm of the moon is accompanied by a subtle, rhythmical motion of the earth: the nutation, or nodding, of the tilted earth's axis. The earth spins on its axis like a top. Like a spinning top, its axis simultaneously performs a slower circular gyrating motion. At present the earth's axis extends toward a point close to the North Star. This point shifts gradually over time as the earth's axis gyrates and describes a complete circle in about 25,770 years. Now, as the earth's axis describes this circle, it nods gently up and down, plunging and rising like a ship moving through the sea. This wavelike motion of the earth's axis, its nutation, also has a periodicity of about 18.6 years (see diagram 3). Thus the earth responds to the greater rhythm of the moon with a simple wavelike motion of its own. This nutation of the earth's axis, however, means that the orientation of the earth to the entire cosmos undergoes constant change in an 18.6 year rhythm. It can be thought of as a rhythmical breathing process in the earth's relationship to the cosmos, which takes place in response to the greater moon rhythm.

Rudolf Steiner characterizes this rhythm in the following way:

> We can therefore indeed perceive the breathing of the macrocosm. We only need to observe the course of the earth, or respectively, the nutation of the earth. The earth dances.... This dance mirrors the breathing of the macrocosm. So that we can say: We look into the breathing of the macrocosm through this movement of the nutation, or of the moon.[49]

We can now distinguish three rhythms of the human entelechy: Each night it is slightly drawn out into the cosmos during sleep, which the ancients called "the little brother of death." At death itself it is drawn out far more powerfully. In the 18.6-year cosmic rhythm just described, the soul—and Goethe's soul in particular—is drawn out to a degree that can resemble a death experience. However, as in sleep or in death, a deeper connection with the cosmos is achieved, and a corresponding inner renewal and transformation takes place.

It is awe-inspiring to observe how Goethe, who experienced such a deep inner connection with nature and who was so keenly aware of the human being's intimate microcosmic relationship to nature's macrocosm, actually lived out his life in such deep accord with this cosmic rhythm. We are reminded of his essay "Nature," quoted on page 35:

> Nature! We are surrounded and embraced by her—powerless to leave her and powerless to enter her more deeply. Unasked and without warning she sweeps us away in the round of her dance and dances on until we fall exhausted from her arms.[50]

Borne by this cosmic breathing rhythm, the deeper, Faustian impulse of Goethe's entelechy was able to unfold on ever new levels. Through its renewing powers Goethe succeeded, over the course of his lifetime, in transforming his "subjective" artistic capacities of imaginative thought, inspired feeling, and intuitive will into organs of "objective" perception. These transformed capacities became the "instruments" of a new, living science through which he was able to participate ever more fully in the creative dimensions of reality. It was in this spirit that Goethe wrote, in conjunction with his study of color:

> Insofar as we make use of our healthy senses [and I would like to add here: *and of our transformed capacities of perceptive thinking, feeling, and willing*] we ourselves are the best and most exact scientific instruments possible. The greatest misfortune of modern physics is that its experiments have been set apart from the human being, as it were; physics refuses to recognize nature in anything not shown by artificial instruments, and even uses this as a measure of its accomplishments.[51]

Our analytical age has isolated human experience from the world of science, the subjective from the objective, the creative spirit from nature. This has led to the fragmentation of factual knowledge, to the isolation and illness of the subjective self, and to the neglect and exploitation of nature. Goethe's life-endeavor can be seen as a great work of synthesis: as an active breathing between the depths of his own inner being and the widths of the outer world, a rhythmical breathing through which he built a living bridge leading the creative human spirit to the creative spirit in nature. Through this cognitive breathing activity he laid the foundations of a science that encompasses, as living idea, the fullness of artistic experience and that leads to a deeply religious view of the world. For, in the light of this science, there is nothing material that is not a revelation of the spirit, and there is no spirit that does not reveal itself in the material world.

At the heart of Goethe's experience lay the consciousness that the human being is a microcosm, the crowning achievement of creation, into whom all the forces of nature have flowed. It was therefore his deep conviction that, by developing the forces and faculties that lie within us, we will again find our way to a living connection with the creative forces in nature, in the macrocosm.

Thus Goethe's innermost impulse can be seen as a healing one: as the impulse to make whole. We find this sense of wholeness beautifully captured in the following words:

> When our healthy human nature works as a totality, when we feel ourselves within the world as within a great, beautiful, wonderful, and precious whole, when our harmonious sense of well-being fills us with pure, uninhibited delight, then the universe—if it were conscious of itself—would exult with joy in having achieved its goal and marvel at the crowning of its own becoming and being.[52]

In conclusion it must be said for the benefit of those for whom this is a first introduction to Goethe's life that this brief sketch can only begin to do justice to its extraordinary richness and depth. It is not for nothing that Goethe is generally considered to be the greatest German of all time by those who know him.[53] Important aspects of his life and work could not be touched on here. Attempting to grasp the fullness of Goethe's genius is almost like trying to grasp that of nature herself.

Today western civilization has reached a crisis. Finding the way from our own human spirit to the creative spirit hidden—yet revealed—in nature and in our fellow human beings has become a life and death question—the only alternative to complete dis-integration. The waters are rising.... Goethe's ship is afloat and ready to sail.

NOTES

1. Goethe, *West-östlicher Divan*, last verse of "Selige Sehnsucht," 1814, translated by J. Barnes.
2. Goethe's essay "Nature," in Suhrkamp, vol. 12, *Scientific Studies* (1987), p. 4.
3. The first lines of a poem by Goethe, "Vom Vater hab' ich die Statur...," translation by John Barnes.
4. Goethe, *Poetry and Truth*, Part II, Book 8, Suhrkamp Edition (1987) vol. 4, p. 248.
5. Herder, J.G., Essay on Shakespeare, in *Von deutscher Art und Kunst.*
6. Friedrich Hiebel, *Goethe, die Erhöhung des Menschen.*
7. Goethe's *Faust* ends with the "Chorus Mysticus," the last lines of which are: "The eternal-feminine draws us onwards."
8. Goethe's essay "Der Verfasser teilt die Geschichte seiner botanischen Studien mit" in *Zur Morphologie.*
9. Goethe, "On Granite," in Suhrkamp, *Scientific Studies*, p. 132-3.
10. Goethe, *Faust I,* lines 382-3, translation by W. Kaufmann.
11. Goethe, "Nature," in Suhrkamp, *Scientific Studies*, p. 3.
12. Oxford Book of English Verse, p. 628-9.

13. Goethe, *Italian Journey*, Suhrkamp, vol. 6, p. 120.
14. Ibid, p. 53-54.
15. Goethe, quoted in *The Metamorphosis of Plants*, published by the Biodynamic Farming and Gardening Association, 1974, p. 16-17.
16. Ibid, p. 17.
17. Goethe, *Italian Journey*, Jan. 6, 1787, Suhrkamp, vol. 6, p. 127.
18. Goethe, *Italian Journey*, Sept. 6, 1787.
19. Goethe's essay "Simple Imitation, Manner, Style," in Suhrkamp, vol. 3, translation by John Barnes.
20. Goethe, *Maxims and Reflections.*
21. Goethe, *Maxims and Reflections.*
22. Goethe, letter to Frau von Stein, July 10, 1786, translation by John Barnes.
23. Goethe, *Conversations with Eckermann*, October 9, translated by J. Oxenford, San Francisco: North Point Press, 1984; passage retranslated by J. Barnes.
24. R.W. Emerson, *Representative Men*, "Goethe; or the Writer."
25. Schiller's letter to Goethe, August 23, 1794.
26. Goethe's introduction to his correspondence with Schiller.
27. Goethe, *Maxims and Reflections*, Suhrkamp, *Scientific Studies*, p. 307.
28. Wyler and Ames, *The Golden Book of Astronomy*, quoted from H.O. Proskauer, *The Rediscovery of Color*, p. 154.
29. Goethe, "Significant Help Given by an Ingenious Turn of Phrase," Suhrkamp, *Scientific Studies*, p. 39.
30. Goethe, *Geschichte der Farbenlehre*, translation by John Barnes.
31. Goethe to Eckermann, Feb. 19, 1829; in Oxenford, p. 245-6.
32. Michael Polanyi, *Meaning*, University of Chicago Press, 1975, p. 63.
33. Goethe, *Maxims and Reflections*, Suhrkamp, *Scientific Studies*, p. 308.
34. Letter to Zelter, April 29, 1830.
35. Goethe, *Maxims and Reflections*, translation by John Barnes.
36. Goethe to Eckermann, Feb. 4, 1829.
37. Goethe, *Faust*, translated by G.M. Priest, Alfred A. Knopf, N.Y., 1957, p. 71.

38. Ibid, p. 188.
39. Rudolf Steiner characterized Faust's three encounters with Helen in his lecture of November 3, 1917.
40. See footnote 37, p. 269.
41. See footnote 2.
42. In his poem "To Charlotte von Stein" (1776) Goethe expressed the intuitive sense that they must have been closely connected in a former life:

    Ach, du warst in abgelebten Zeiten
    Meine Schwester oder meine Frau.
    (O, you were in once-lived times
    My sister or my wife.)

    In a conversation with Falk, Goethe said: "As you see me here, I have surely existed a thousand times before and expect to return a thousand times more." Biedermann, 1490.
43. From a letter written by Goethe in his mid-thirties quoted in E. Ludwig, *Goethe, the History of a Man*, p. 206.
44. Goethe, *William Meister's Journeyman's Years*, translation by John Barnes.
45. Goethe, from a conversation with B. Constant on January 23, 1804; Biedermann 733; translation by John Barnes.
46. Goethe to Eckermann, March 11, 1828.
47. This poem is from the *West-östlicher Divan*: "Im Atemholen..," translation by John Barnes.
48. The New International Encyclopedia, vol. IV, 1923, p. 224.
49. Rudolf Steiner, lecture of April 16, 1920, translation by John Barnes.
50. See footnote 11.
51. Goethe, *Maxims and Reflections*, Suhrkamp, *Scientific Studies*, p. 311.
52. Goethe, from Goethe's essay on Winckelmann, in Kürschner's National Literature, vol. 27, p. 42, translation by John Barnes.
53. W. Kaufmann, *Goethe's Faust*, Anchor Books, 1963, Introduction, p. 3.

BIBLIOGRAPHY

Baumgaertner, A. 1911. *Goethe, sein Leben and seine Werke.* Freiburg.

Biedermann, F. 1909-1911. *Goethes Gespräche.* Leipzig.

Bielschowsky, A. 1919. *Goethe, Sein Leben und seine Werke,* Munich: C.H. Beck'sche Verlagsbuchhandlung.

Boyle, N. 1991. *Goethe: The Poet and the Age. Volume I, The Poetry of Desire.* New York: Oxford University Press.

Friedenthal, R. 1963. *Goethe, Sein Leben und seine Zeit.* Munich.

Goethe, J.W. 1988. Suhrkamp Edition in 12 volumes. New York.

Hahn, H. 1966. *Der Lebenslauf als Kunstwerk.* Stuttgart: Verlag Freies Geistesleben.

Hiebel, F. 1961. *Goethe, die Erhöhung des Menschen,* Bern: Francke Verlag.

Ludwig, E. 1928. *Goethe, the History of a Man.* New York: Putnam's Sons.

Meyer, R. 1949. *Goethes Lebensstufen.* Schaffhausen: Columban Verlag.

Mueller, G. 1947. *Kleine Goethebiographie.* Frankfurt: Athenaeum Verlag.

Steiner, R. Many of his writings and lectures on Goethe, particularly the first two lectures in the series "The Karma of Vocation" given in Dornach on November 4 and 5, 1916.